Collins · *do brilliantly!*

Instant**Facts**

Chemistry

A-Z of **essential facts** and definitions

William A. H. Scott

William Collins' dream of knowledge for all began with the publication of his first book in 1819. A self-educated mill worker, he not only enriched millions of lives, but also founded a flourishing publishing house. Today, staying true to this spirit, Collins books are packed with inspiration, innovation and practical expertise. They place you at the centre of a world of possibility and give you exactly what you need to explore it.

Collins. Do more.

Published by Collins
An imprint of HarperCollins*Publishers*
77–85 Fulham Palace Road
Hammersmith
London
W6 8JB

Browse the complete Collins catalogue at

www.collinseducation.com
© HarperCollins*Publishers* Limited 2005

First published as Collins Gem Basic Facts Chemistry 1982

10 9 8 7 6 5 4 3 2 1

ISBN 0 00 720514 7

British Library Cataloguing in Publication Data
A catalogue record for this publication is available from the British Library

Every effort has been made to contact the holders if copyright material, but if any have been inadvertently overlooked, the Publishers will be pleased to make the necessary arrangements at the first opportunity.

Edited and Project Managed by Marie Insall
Production by Katie Butler
Design by Sally Boothroyd/Jerry Fowler
Printed and bound by Printing Express, Hong Kong

You might also like to visit
www.harpercollins.co.uk
The book lover's website

Introduction

Instant Facts Chemistry is one of a series of illustrated A–Z subject
reference guides of the key terms and concepts used in the most important
school subjects. With its alphabetical arrangement, the book is designed
for quick reference to explain the meaning of words used in the subject
and so is an excellent companion both to course work and during revision.

Bold words in an entry identify key terms which are explained in
greater detail in entries of their own; important terms that do not have
separate entries are shown in *italic* and are explained in the entry in
which they occur.

Elements are presented in this consistent format:

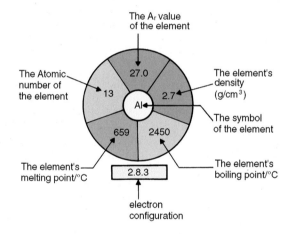

The A_r value
of the element

The Atomic
number of
the element

The element's
density
(g/cm^3)

The symbol
of the element

The element's
melting point/°C

The element's
boiling point/°C

electron
configuration

Other titles in the *Instant Facts* series include:
English
Modern World History
Biology
Science
Physics
Geography
Maths

A

A_r The symbol for **relative atomic mass**.

absolute temperature *See* **Kelvin temperature scale**.

abundance A measure of how much of a substance exists; for example, how much of each element there is in a planet or of **isotopes** in an element. It is often expressed as a percentage.

(a) Earth

elements	O	Si	Al	Fe	Ca	Na	K	Mg
% abundance	46.6	27.7	8.1	5.0	3.6	2.8	2.6	2.1

(b) Sun

elements	H	He	O	C	Si
% abundance	54.0	44.7	0.8	0.4	0.05

abundance *The abundance of (a) the elements in the Earth and (b) the elements in the Sun.*

Element	Isotope	% abundance
Chlorine	$^{35}_{17}Cl$	75.5
	$^{37}_{17}Cl$	24.5
Carbon	$^{12}_{6}C$	98.9
	$^{13}_{6}C$	1.1
Magnesium	$^{24}_{12}Mg$	78.6
	$^{25}_{12}Mg$	10.1
	$^{26}_{12}Mg$	11.3

abundance *The isotopes of some commonly occurring elements.*

Knowing the percentage abundance of each isotope it is possible to calculate the **relative atomic mass** (A_r) for the element, e.g. for chlorine:

$$^{35}_{17}Cl : ^{37}_{17}Cl = 3:1$$

$$A_r(Cl) = \frac{(3 \times 35) + (1 \times 37)}{4} = 35.5$$

accelerator Any chemical used to speed up **cross-linking** reactions in **polymers** or the **curing** of **epoxy resins**. *See also* **catalyst**.

accumulator A rechargeable **battery**. The most common type is the lead–acid accumulator. This is the battery that is used to start cars and lorries and to drive milk floats, fork-lift trucks and experimental electric cars. In the lead–acid accumulator the **electrolyte** is dilute sulphuric acid. The positive plate is made of lead(IV) oxide and the negative plate is composed of lead. When the accumulator is discharged (flat) the plates are coated with lead(II) sulphate.

The sulphate coating is removed when the accumulator is recharged by passing electricity through it. The lead-acid accumulator is dangerous because it can produce very high currents. It produces the flammable gas hydrogen whilst it is being recharged and there is the constant danger of the acid spilling from the accumulator.

acetic acid *See* **ethanoic acid**.

acetylene *See* **ethyne**.

acid Any substance that releases **hydrogen ions** (H⁺) when it is added to water. The hydrogen ion is *solvated*, that is, a water molecule adds on to it, to give the **oxonium ion** (H_3O^+). Acidic solutions have a **pH** of less than 7, and have an excess of H⁺ ions in solution. Common laboratory acids include nitric acid (HNO_3), hydrochloric acid (HCl) and sulphuric acid (H_2SO_4).

These acids are dangerous, corrosive liquids and should always be treated with care.

Acids have the following properties:
(a) they turn blue **litmus** red;
(b) they emit carbon dioxide when added to **carbonates**;
(c) they give off hydrogen when added to certain metals;
(d) they neutralize **alkalis**.
See also **strength of acids and bases**.

acid–base reaction A reaction between an **acid** and a **base** to form a **salt** and water only. For example:

Hydrochloric +	Sodium	→	Sodium	+	water
acid	hydroxide		chloride		$H_2O(l)$
HCl(aq)	NaOH(aq)		NaCl(aq)		

acidic oxides **Oxides** of non-metals that react with water to form acidic solutions. Examples are given in the table.

Oxide		Acid	
Carbon dioxide	CO_2	Carbonic acid	H_2CO_3
Sulphur dioxide	SO_2	Sulphurous acid	H_2SO_3
Sulphur trioxide	SO_3	Sulphuric acid	H_2SO_4

acidic oxides *Examples of acids and the oxides from which they derive.*

acidification   The fall in **pH** of water in lakes, rivers, wells and in the ground, caused by pollutants such as sulphur dioxide (SO_2) and nitrogen oxides (NO, N_2O and NO_2). These pollutants are produced in power stations, industry, the home and internal combustion engines. They rise up into the atmosphere where they react to form acids, which can be carried thousands of kilometres before falling to the ground, as **acid rain**, causing serious pollution. Animal and plant life are affected, as are buildings made of stone.

acid rain   Rain polluted by a build-up of acids in the atmosphere. Rain is naturally acidic because of the carbon dioxide which is dissolved in it. However, hydrogen chloride and sulphur dioxide produced by industrial and some natural activities (such as earthquakes) combine with oxygen and water vapour in the atmosphere to produce strong acid solutions. These make the rain more acid, which causes much damage to trees, lakes and buildings. The normal **pH** of rain is 5.6. The lowest pH of rain recorded in the UK is 2.4, which is over 1000 times more acidic than normal.
See also **acidification, pollution.**

acid salts   **Salts** in which only some of the replaceable **hydrogen** atoms in an **acid** molecule have been replaced by a metal. For example if only one of the hydrogen atoms in carbonic acid, H_2CO_3, is replaced by a sodium atom the acid salt sodium hydrogencarbonate, $NaHCO_3$, is formed. If both hydrogens are replaced the *normal salt* sodium carbonate, Na_2CO_3, is formed.

actinide elements   *See* **periodic table.**

activation energy (E_a)
The energy needed to start off a chemical **reaction**. When hydrogen and oxygen are mixed there is no reaction. However, when a flame is brought into contact with the gases, there is an immediate explosion.

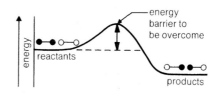

activation energy *The energy barrier that has to be overcome for a reaction to take place.*

The heat of the flame causes the reaction to occur. Many chemical reactions are like this. It seems that there is a barrier that has to be overcome before the reaction will take place. Energy, as in the heat of the flame, has to be put in to make the reaction occur. *See also* **catalyst**.

active carbon *See* **charcoal**.

activity series *See* **electrochemical series**.

addition polymer A **polymer** made by an **addition reaction**.

addition reaction A reaction in which two or more molecules are reacted together to form a single molecule. Good examples are found in the reactions of **alkenes**. These compounds contain **double bonds** and can add an atom to each side of the double bond to form a saturated **alkane**. Addition reactions also occur in the making of many **polymers**, e.g. **poly(ethene)**, where many thousands of ethene molecules join to form a long chain.

ethene + hydrogen ⟶ ethane

addition reaction *Ethene reacting with hydrogen.*

additive A small amount of substance added to a material to give it particular properties. Additives are widely used today. Examples include:
- anti-foaming agents in washing powders;
- colouring materials in food and soft drinks;
- **emulsifiers** in margarine.

adhesive A substance used to stick one material to another. Examples include glues, plant resins and **epoxy resins**. Adhesives are **polymers**. Some are dissolved in a solvent which, when used, evaporates and leaves the adhesive behind to hold materials together.

aerosol A **colloidal dispersion** in which very small liquid or solid particles are suspended in a gas. Natural examples include fog, clouds and smoke. The word is most commonly used, however, for the device which is used to create aerosols, the **aerosol can**.

aerosol can A device for providing small amounts of chemicals in a finely divided form. Aerosol cans have been used to deliver anti-perspirants, shaving foams, paint, lubricants, ointments and foods such as cream. Because of the concern over **CFC** gases, the propellant used to push the contents out of the can is now usually an **alkane** such as butane. This is normally liquefied under pressure and mixes with the material which is to be delivered. *See also* **aerosol**.

aggregate Any material that can be cemented (bound) together to form a solid material. Aggregates are used in road surfaces, in the foundations of buildings and in producing building materials such as bricks and concrete.

agrochemicals Natural and synthetic chemicals used in the agriculture industry. Examples include **biocides**, growth regulators, soil conditioners, and vitamin and mineral supplements. Note that **fertilizers** are not included in this definition.

air The mixture of gases that surrounds the Earth. The average composition of pure air is shown below. This composition varies from place to place and also varies with altitude. *See also* **Earth's atmosphere**.

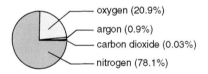

oxygen (20.9%)
argon (0.9%)
carbon dioxide (0.03%)
nitrogen (78.1%)

air *The composition of air. Other components include neon, helium, krypton, xenon (in small constant amounts) water (in very variable amounts).*

Air also contains pollutants, some of which are shown in the table.

Pollutant	Sources
Sulphur dioxide	Burning coal, oil
Carbon monoxide	Engines, cigarettes
Oxides of nitrogen	Car engines
Soot	Fires, engines
Pollen	Trees, flowers
Dust	Volcanoes

air *Some pollutants.*

Air is vital for life. The oxygen is necessary for **respiration** and the carbon dioxide is needed for **photosynthesis**. *See also* **pollution**.

alcohols Important **organic** compounds. They possess the arrangement of atoms:
The –OH group is the **functional group** of the alcohols. **Ethanol** is the most important compound but *see also* **methanol** and **glycol**.

Alcohols with a small **relative molecular mass** are flammable liquids which dissolve in water. Some of the important properties of alcohols are that they can be oxidized easily (*see* **oxidation**), they form **esters** and react with sodium to produce hydrogen.

alkali Any **base** that is soluble in water. Alkalis are usually metal **hydroxides**, e.g. sodium hydroxide, but ammonia solution is also an alkali. Their reactions are affected by their strength and their concentration. Oven cleaners, household ammonia and some paint strippers contain alkalis. Alkalis have an excess of OH–ions in solution and have the following properties:
(a) they turn red **litmus** blue;
(b) they neutralize **acids**;
(c) they have a **pH** above 7;
(d) they react with acids to produce a **salt** and water only.

alkali metals Very reactive metals found in group I of the **periodic table**. These metals react with water to form **alkaline** solutions. For example:

$$2Na(s) + 2H_2O(l) \rightarrow 2NaOH(aq) + H_2(g)$$

alkaline (used of a solution) Having a **pH** greater than 7. *See also* **alkali**, **alkali metals**, **alkaline earth metal**.

alkaline earth metal A metal found in group II of the **periodic table**. Such metals are less reactive than the **alkali metals** found in group i, but, like them, produce **alkaline** solutions when reacted with water.

alkanes **Hydrocarbon** compounds that are important in our lives. The chief source of alkanes is **petroleum**. They are valuable **raw materials** in the chemical industry, and domestic gas supplies are almost 100% alkanes. Vaseline, or petroleum jelly, is also made of alkanes. They have a **general formula** C_nH_{2n+2} so the alkane with 4 carbon atoms will have 10 hydrogen atoms: C_4H_{10}, **butane**.

Alkanes are **saturated** compounds and so are not very reactive. They tend to be flammable and will react with chlorine in the presence of **ultraviolet radiation**. *See also* **methane, ethane, propane**.

Alkanes may also be cyclic (*cycloalkanes*) instead of linear, with a general formula C_nH_{2n}. Cyclohexane (C_6H_{12}) is a common example. *See also* **cyclic compounds**.

alkenes Important **hydrocarbon** compounds. They are widely made in oil refineries and are used as starting materials in the manufacture of many materials, including **plastics. Ethene**, **propene** and styrene (phenyl-ethene) are three of the most important alkenes. They have a **general formula** C_nH_{2n}; the alkene with 3 carbon atoms (propene) will therefore have 6 hydrogen atoms. Because they are **unsaturated compounds** they have a carbon–carbon **double bond** and are reactive compounds. Typically they will undergo **addition reactions** with hydrogen, **halogens** and water. They will also form **polymers** by this addition process. *See also* **poly(phenylethene)** and **poly(propene)**.

alkynes Hydrocarbons with the **general formula** C_nH_{2n-2}; **ethyne** with 2 carbon atoms will have 2 hydrogen atoms: C_2H_2. Alkynes have carbon–carbon **triple bonds** and therefore show the typical **addition reactions** of **unsaturated compounds**.

allotropes Elements that exist in different forms in the same physical state. The chemical properties are the same but the physical properties are different. The best example is the element carbon, which has two common allotropes: **diamond** and **graphite**. Diamond is hard and colourless, graphite is flaky and black. Other elements that exist in different allotropic forms include sulphur, phosphorus and tin.

alloy A **mixture** which is made up of two or more **metals** or which contains metals and non-metals. Alloys are much more widely used than pure metals because, by bringing two or more elements together in the right proportions, specific properties can be obtained. For example, aluminium is quite a soft metal but when a small amount of copper is mixed in, the alloy duralumin is produced, which is strong enough to be used in aircraft frames. Some common examples are **brass**, **bronze**, **pewter** and **steel**.

alpha particles ($^4_2$He) Helium atoms without their electrons. They are produced in many **nuclear reactions**. They have a fairly short range in air (usually less than 10 cm) and are easily stopped by thin sheets of paper or foil.

alumina Another name for aluminium oxide (Al_2O_3). It is the chief constituent of **bauxite** from which **aluminium** is produced. Alumina itself is useful as a **refractory material** and as a support material for **catalysts**.

aluminium (Al) The most abundant metal in the Earth's crust (approximately 8% by mass). Clay, shale, slate and granite all contain aluminium compounds but the metal is difficult to extract from them. It is obtained by the **electrolysis** of **bauxite** dissolved in **cryolite**, using graphite electrodes in the **Hall–Hérault cell**.

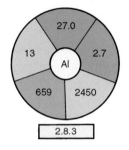

aluminium

The metal has a **valency** of 3. Although it is very reactive, aluminium and its alloys are resistant to **corrosion** because a protective layer of **oxide** forms on the surface of the metal when it is exposed to air. It is subject to anodizing (*see* **anodize**).

Aluminium is **malleable**, easy to handle and has a low density, and its alloys can be very strong. The metal and alloys are used in a great many ways. Examples include:
(a) construction, in window and door frames;
(b) transport, in car bodies, engines, aeroplanes, cycle frames, ship superstructures;
(c) the electricity industry, in power lines;
(d) the food industry, in milk bottle tops and baking foil.

aluminium compounds

Aluminium chloride $AlCl_3$	The anhydrous chloride is covalent.
Aluminium oxide Al_2O_3	The oxide is amphoteric.
Aluminium sulphate $Al_2(SO_4)_3$	The most important aluminium compound. It is used as a precipitator in sewage works, as a mordant and as a size in the paper industry. It is also used as a foaming agent in fire extinguishers.

amalgam Any **alloy** that contains mercury. Zinc amalgam is used for teeth fillings. *Amalgamation*, the process of forming an amalgam, was once used to extract gold and silver from crushed rock which contained the metals.

amino acids Organic compounds that contain –COOH and –NH$_2$ groups. They are the building bricks from which **proteins** are made. Twenty different amino acids are needed to make up all the proteins in our bodies. Our bodies can produce some of these amino acids but we need to take in eight particular ones through eating. These are known as *essential amino acids*. Amino acids combine together to form **peptides**.

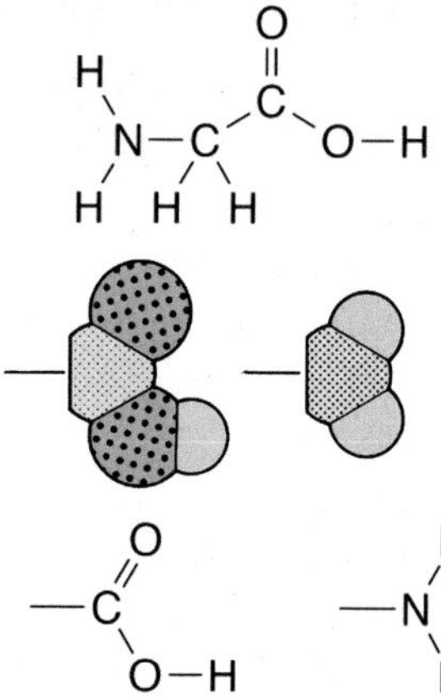

ammonia (NH$_3$) A colourless gas with a pungent odour. It is very soluble in water giving an **alkaline** solution. Ammonia was once made from coal but now over 90% is produced by the **Haber process**.

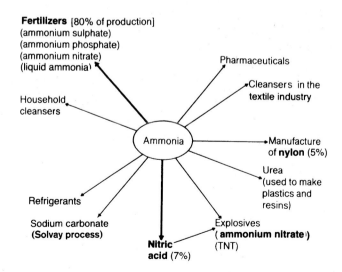

amino acids *Glycine, the simplest of the amino acids.*

Ammonia is also produced by bacteria found on the roots of leguminous plants (peas and beans). When protein decomposes, ammonia is released. Both of these are important sources of plant food. *See also* **nitrogen cycle**.

Fertilizers [80% of production]
(ammonium sulphate)
(ammonium phosphate)
(ammonium nitrate)
(liquid ammonia)

Pharmaceuticals

Cleansers in the textile industry

Household cleansers

Ammonia

Manufacture of **nylon** (5%)

Urea (used to make plastics and resins)

Refrigerants

Sodium carbonate
(**Solvay process**)

Explosives
(**ammonium nitrate**)
(TNT)

Nitric acid (7%)

ammonia *The uses of ammonia.*

Ammonia is a **covalent compound**. It has a characteristic reaction with hydrogen chloride to give dense white fumes of ammonium chloride:

$$NH_3(g) + HCl(g) \rightarrow NH_4Cl(s)$$

Ammonia can be oxidized (*see* **oxidation**) to nitric acid and this is the major source of nitric acid today. It is easily liquefied and it is usually carried in this state from place to place in tankers.

ammonia solution (NH_4OH)   (formerly called ammonium hydroxide) A solution of **ammonia** in water.

ammonium compounds

Ammonium nitrate NH_4NO_3	A very important compound that is used as a fertilizer and an explosive.
Ammonium phosphate $(NH_4)_3PO_4$	A convenient chemical with which to put both nitrogen and phosphorus into the soil.
Ammonium sulphate $(NH_4)_2SO_4$	An important fertilizer.

ammonium ion   An ion found in ammonium compounds as well as in a solution of ammonia gas in water. The ion is positively charged and the hydrogen atoms are arranged around the nitrogen atom in a tetrahedral way (*see* **tetrahedral compound**).

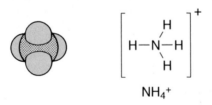

$$\left[\begin{array}{c} H \\ \diagup \\ H-N-H \\ \diagdown \\ H \end{array} \right]^{+}$$

$$NH_4^+$$

ammonium ion

amorphous   Without definite shape or form. Any substance that is non-crystalline, that is, it appears to have no regularity of shape, is described as amorphous.

ampere (amp)   The unit of electric current. Its symbol is A. It measures the rate of flow of charge. 1 A = 1 **coulomb/second**. *See also* **SI units**.

amphoteric (relating to insoluble **oxides** and **hydroxides**) Showing both basic and acidic properties. Amphoteric compounds, such as zinc oxide (ZnO), will react with both **acids** and **alkalis** to form **salts**:

$$ZnO + 2HCl \rightarrow ZnCl_2 + 2H_2O$$

$$ZnO + 2NaOH + H_2O \rightarrow Na_2Zn(OH)_4$$
(sodium zincate)

amu   *See* **atomic mass unit**.

anaemia   An illness in which blood does not contain sufficient iron.

anaesthetic   A substance that reduces or removes the feeling of pain. General anaesthetics, such as **fluothane** (C_2HF_3BrCl), produce deep unconsciousness. Local anaesthetics only affect a small part of the body while the person stays awake. *See also* **ethers**, **nitrogen oxides**.

analysis   The investigation or testing of substances to find out what they are or what more elementary substances they contain. **Qualitative** analysis provides information about what the material contains; for example, which elements are present in a compound or which compounds are found in a mixture. **Quantitative** analysis provides information about how much of particular substances there is; for example, the percentage composition of a food in terms of fat, protein and carbohydrate, or the amount of alcohol in a person's blood.

anhydrides   Substances that react with water to produce **acids**. All **acidic oxides** are anhydrides, but some will give rise to one acid only whilst others will produce two. For example, SO_3 gives rise to H_2SO_4; NO_2 gives rise to HNO_3 and HNO_2.

anhydrite   The mineral calcium sulphate ($CaSO_4$). It is widely used in the manufacture of **sulphuric acid**.

anhydrous   Containing no water. The term usually denotes **salts** with no **water of crystallization**. Examples of anhydrous salts are anhydrous copper(II) sulphate $CuSO_4$ and anhydrous sodium carbonate Na_2CO_3. The corresponding **hydrated** salts are $CuSO_4.5H_2O$ and $Na_2CO_3.10H_2O$.
   The term 'anhydrous' is also used to describe liquids which are perfectly dry, i.e. that contain no water, for example anhydrous **ether**.

anion   A negatively charged **ion**. Anions usually contain a full outer orbital of **electrons**. Anions are attracted to the **anode** in **electrolysis**. Non-metals form anions but complex ions can also be anions.

annealing A method of strengthening a material by heating and controlled cooling. Heating and cooling metals causes changes in their properties, especially their strength. When a metal is heated to a high temperature and then cooled slowly the crystals grow very large. The metal then becomes **malleable**.

The continuous annealing process now used in the production of **tinplate** is much more rapid than the previous batch process. What used to take many hours now takes only a few minutes.

anode The **electrode** that carries the positive charge in a solution that is undergoing **electrolysis**. **Anions** are attracted to the anode because they are negatively charged. These **ions** give up their extra electrons at the anode. Nonmetallic elements are produced and the electrons travel round the circuit. *Compare* **cathode**.

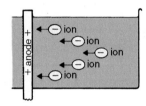

anode *Negatively charged ions attracted to the positive anode.*

anodize To coat a metal with a protective **oxide** film. For example, aluminium has a protective layer of oxide (Al_2O_3). This layer of oxide can be thickened by making the piece of aluminium the **anode** (+) in an electrolytic **cell**. Dilute sulphuric acid is used as the **electrolyte**. Aluminium atoms give up electrons and react with the water atoms:

$$2Al(s) + 3H_2O(l) \rightarrow Al_2O_3(s) + 6H^+(aq) + 6e^-$$

Aluminium oxide is formed and this can be dyed to produce attractive finishes to products, such as coloured milk-bottle tops.

antacids Substances used to increase the **pH** of stomach juices and, therefore, relieve indigestion. Sodium hydrogencarbonate ($NaHCO_3$) and magnesium oxide (MgO) are common examples.

antifreeze A substance that is added to the cooling system of engines in winter to prevent the formation of ice which would damage the engine. The added substances lower the freezing point of the water. Examples are methanol and ethane-1,2-diol.

antioxidant Natural substances or chemical **additives** that slow down oxidation reactions. They are widely added to manufactured foods, particularly those containing fat. They increase the shelf-life of the food, that is, the amount of time it can be kept before being eaten. Many antioxidants found in food are natural substances. **Vitamins** A, C and E are good examples.

Antioxidants, usually in the form of **aromatic** compounds, are also added to a wide number of **rubber** and **plastic** materials to prevent oxidation which would result in a loss of function of the material.

aq Abbreviation used to denote an **aqueous solution**, e.g. NaCl(aq).

aqueous solution A **solution** where water is the **solvent**.

argon (Ar) The most abundant **noble gas**. It makes up 0.9% of the atmosphere (by volume). It is very unreactive. No compound of argon has yet been made. Argon has uses where the presence of an unreactive gas is needed. 90% of argon production is used in arc welding and in metal manufacture, melting and casting; 10% is used, mixed with a small amount of nitrogen, in electric light bulbs and fluorescent tubes.

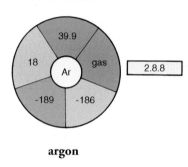

argon

aromatic (used of compounds) Containing a **benzene** ring structure in their molecules.

arsenic (As) A brittle grey **metalloid**. Its compounds are very poisonous – especially the oxide (As_2O_3). It is a group V element that has three **allotropes** – yellow, black and grey. Arsenic compounds are used in pesticides. The grey allotrope sublimes (see **sublimation**) at 613 °C.

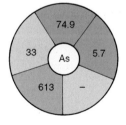

arsenic

asbestos A naturally occurring aluminosilicate (a compound containing aluminium and silicon). It was once widely used for fireproof clothing and insulating materials, but its use today is restricted because of the dangers of contracting cancer of the lung lining and *asbestosis*, a fatal lung disease.

asphalt   A mixture of **bitumen** and **aggregate** used widely in the surfacing of roads.

atmosphere   **1.** A unit of **pressure**. Although the pressure of the air that surrounds us varies from place to place and from time to time, its value usually only varies slightly and is always approximately one atmosphere.
   2. The mixture of gases that surrounds a planet, sun or moon. *See* **Earth's atmosphere**.

atom   The smallest indivisible particle of an **element** that can exist, the building bricks with which everything is made. An atom can be thought of as the smallest part of an element that can take part in a chemical **reaction**. They are small particles; 100 million placed end-to-end would measure 1 cm. They are made up of even smaller **subatomic particles**:
- **neutrons** (n) and **protons** (p) are particles found in the centre of the atom – the **nucleus**;
- **electrons** (e) are particles that orbit the nucleus.
   The atom as a whole is electrically neutral although the protons and electrons carry electrical charges. These charges are equal in size but opposite in sign, hence the number of protons always equals the number of electrons. Atoms that have lost or gained electrons are called **ions**. All atoms of the same element contain the same number of protons and therefore have the same **atomic number**, but atoms of the same element can contain different numbers of neutrons. *See also* **isotope**.

atomic mass unit (amu)   A unit used to express the masses of **nuclei** and **nucleons**, equal to exactly one twelfth the mass of a neutral atom of the carbon-12 **isotope**. The atomic mass unit has a value of 1.660×10^{-27} kg. The **proton** and the **neutron** are each given the value of 1 amu.
   It is not usual to refer to the mass of an atom in grams or kilograms so the mass of one atom is normally compared with a standard mass. This is called the **relative atomic mass** (A_r).

atomic number (Z)   The number of **protons** in the **nucleus** of an **atom**. All atoms of the same **element** have the same atomic number, e.g. sodium atoms all contain 11 protons. In a **neutral** atom the number of **electrons** equals the atomic number.

atomicity   The number of **atoms** in a **molecule** of an **element**. With the **noble gases** there is only one atom in the molecule and the atomicity is 1. All other gases have an atomicity of 2 (e.g. oxygen (O_2), nitrogen (N_2)) except ozone (O_3).

Avogadro constant or Avogadro number (L)   The number of **atoms** that are contained in exactly 12g of the carbon-12 **isotope**. More generally it is the number of particles present in a **mole** of substance. $L = 6 \times 10^{23}$ mol^{-1}. The constant is named after the Italian physicist Amedeo Avogadro, 1776–1836, famous for his work on gases.

Avogadro's hypothesis or Avogadro's law   A hypothesis that states that under the same conditions of temperature and pressure, equal volumes of gases contain the same number of molecules, e.g. 10 dm^3 of oxygen and 10 dm^3 of hydrogen both contain the same number of molecules. They contain twice as many molecules as 5 dm^3 of chlorine, provided that all the volumes were measured at the same temperature and pressure.

B

back reaction *See* **equilibrium**.

baking powder A mixture of a **carbonate** and a weak acid that is used in cooking. When water is added to the solid mixture or the mixture is heated, carbon dioxide is produced. This produces bubbles in the dough or cake mixture and it rises, producing a product with an open texture. The mixture usually contains sodium hydrogencarbonate ($NaHCO_3$) and tartaric acid.

balance 1. A device for comparing the **masses** of objects.
2. To have equal numbers of each atom on each side of a chemical equation. This is necessary as atoms can be neither created nor destroyed. For example the following equation is not balanced:

$$Zn(s) + HCl(aq) \rightarrow ZnCl_2(aq) + H_2(g)$$

Number of atoms:

$$Zn = 1H = 1Cl = 1 \text{ (on left)}$$

$$Zn = 1Cl = 2H = 2 \text{ (on right)}$$

To balance the equation there should be equal numbers of atoms on each side of the equation:

$$Zn(s) + 2HCl(aq) \rightarrow ZnCl_2(aq) + H_2(g)$$

The equation is now balanced.

barium A very reactive group II element. It occurs in nature as the **sulphate** (*barytes*) and as the **carbonate**. The metal is extracted from the molten **chloride** by **electrolysis**. Barium chloride solution is used to test for sulphates:

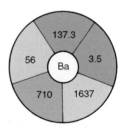

$$Ba^{2+}(aq) + SO_4^{2-}(aq) \rightarrow BaSO_4(s)$$

barium

This equation shows that a **precipitate** of barium sulphate is formed when barium chloride is mixed with a soluble metal sulphate.

barrel **1.** A cask for holding beer. It has a **volume** of exactly 32 imperial gallons.
2. A unit volume in the **petroleum** industry. One barrel = approximately 159 litres (35 imperial gallons).

base A substance that reacts with an **acid** to form a **salt** and water only. Bases are usually metal **oxides** or **hydroxides**, e.g. sodium hydroxide (NaOH) and copper(II) oxide (CuO).

$$NaOH(aq) + HCl(aq) \rightarrow NaCl(aq) + H_2O(l)$$

$$CuO(s) + H_2SO_4(aq) \rightarrow CuSO_4(aq) + H_2O(l)$$

Metal oxides and hydroxides that are soluble in water are known as **alkalis**. These are compounds of group I and II metals.
Ammonia solution is an alkali. It contains **hydroxide ions** in **equilibrium**.

$$NH_3(g) + H_2O(l) \rightleftharpoons NH^+_4(aq) + OH^-(aq)$$

basic oxide An oxide that reacts with an **acid** to form a **salt** and water only. Basic oxides are oxides of metals, but not all metals give basic oxides; for example, aluminium oxide is **amphoteric**. *See also* **acidic oxide**.

basic oxygen process *See* **steel manufacture**.

basic salts **Salts** that contain **hydroxide ions** as well as normal **anions**, e.g. **sulphate**, **carbonate**. Common examples are malachite, $CuCO_3.Cu(OH)_2$, and white lead, $2PbCO_3.Pb(OH)_2$.

batch process An industrial process in which a product is made in a non-continuous way.

battery A number of electric **cells** joined together. Batteries are of two types – either rechargeable, some of which are also known as **accumulators**, or non-rechargeable.
Batteries are available in different shapes, sizes and price ranges and their chemical composition varies too. The common car battery, the lead–acid accumulator, contains six cells, each of which gives 2 volts. The total voltage is 12 volts as the cells are connected together in series in the accumulator.
The common *dry battery* is a single cell. There are four types of dry battery:
(a) zinc–carbon (1.5V)
(b) alkaline manganese (1.5V)
(c) silver oxide (1.5V)
(d) nickel–cadmium (1.4V).

Nickel–cadmium cells give a large current and are rechargeable. Silver oxide cells, used in watches, calculators and cameras, are small and lightweight. The current remains steady when alkaline manganese cells are used, which is useful in cassette players, radios, etc. The zinc–carbon cell is low cost, but has a short life.

bauxite   The chief ore from which **aluminium** is extracted. It is a **hydrated** oxide ($Al_2O_3.xH_2O$) and is found in tropical regions of the world, e.g. Northern Australia and West Africa.

Benedict's solution   A **solution** that can be used as a test for a **reducing agent**. It consists of a mixture of copper(II) sulphate, sodium carbonate and sodium citrate dissolved in water. A brown/red/yellow colour is formed when a reducing agent is warmed with the solution. It is most often used to test for reducing **sugars**, i.e. sugars that are also reducing agents, such as **glucose**.

benzene (C_6H_6)   The simplest **aromatic compound**. It is a toxic liquid **hydrocarbon** which can cause cancers. Its widespread use in schools has been replaced by that of **methylbenzene**. It is produced from **naphtha**, and is an important source of **organic** compounds which are used to produce **poly(phenylethene)**, **phenol**, **detergents** and **nylon**.

benzene

beta particles   High-speed electrons that are produced in the following **nuclear reaction**:

$$^{1}_{0}n \rightarrow {}^{1}_{1}p + {}^{0}_{-1}e$$

For example,

$$^{31}_{15}P + {}^{1}_{0}n \rightarrow {}^{32}_{15}P \rightarrow {}^{32}_{16}S + {}^{0}_{-1}e$$

In this **radioactive** decay the electrons are expelled from the **nucleus** of the **atom**. They travel at high speeds (up to 98% of the speed of light) and have a greater penetrating power than **alpha particles**. When this reaction occurs, the atom concerned changes its **atomic number** because it gains a **proton**.

biocide   A chemical that is used to kill or control living organisms. Biocides used on plants are known as *herbicides*; those used on fungi are called *fungicides*, and those used on animals are known as *pesticides*.

biodegradable Capable of being broken down by microorganisms in the soil. *See also* **biopolymer, photodegradable.**

biogas The gas (about 50% methane) that is produced by the decay of organic waste in the absence of air.

biomass **1.** Any plant material which might be used as a resource. Examples are trees, grass, waste vegetables, sewage, farm slurries. Biomass can be used as a direct source of energy (e.g. by burning wood or straw), and can be decomposed to produce fuels (e.g. by **fermentation** of waste to make **methane**) or as a feedstock for the chemical industry.
  2. The total mass of organisms in a given environment. It is usually expressed as the amount of dry matter per unit area of the environment in which the organisms live.

biopolymer An organic **polymer** produced by bacteria. Biopolymers are biodegradable, and are useful when a material is only needed for a short time, such as for surgical stitches.

bitumen A mixture of **hydrocarbons** of a high boiling point that forms the final residue in the **distillation** of **petroleum**. Bitumen is used for roofing and road surfacing. *See also* **asphalt.**

blast furnace A furnace allowing a continuous production of molten **iron**. At the bottom of the furnace the hot air reacts with **coke** to produce carbon dioxide:

$$C(s) + O_2(g) \rightarrow CO_2(g)$$

This **exothermic** reaction raises the temperature of the furnace to 1800°C. The carbon dioxide then reacts with further coke:

$$C(s) + CO_2(g) \rightarrow 2CO(g)$$

The carbon monoxide so formed reduces the iron **oxides** to iron (1200°C). For example:

$$Fe_2O_3(s) + 3CO(g) \rightarrow 3CO_2(g) + 2Fe(l)$$

The molten iron flows to the bottom of the furnace. The **limestone** in the charge is decomposed by the heat producing calcium oxide and carbon dioxide.
  The calcium oxide reacts with impurities from the iron ore (mainly silica, SiO_2) and forms a molten slag:

$$CaO(s) + SiO_2(s) \rightarrow CaSiO_3(l)$$

This sinks to the bottom of the furnace where it floats on top of the iron. The iron which is produced is called *cast iron*; it contains about 3% carbon and is a brittle metal. In **steel manufacture** the amount of impurities in the iron is very carefully controlled.

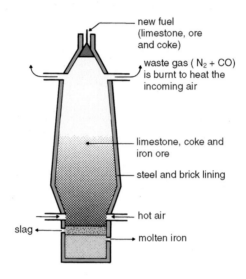

new fuel
(limestone, ore
and coke)

waste gas (N_2 + CO)
is burnt to heat the
incoming air

limestone, coke and
iron ore

steel and brick lining

hot air

slag

molten iron

blast furnace

bleach   A substance used to decolorize materials, e.g. fabrics and paper. Sunlight and oxygen act as bleaches but the most commonly found bleach is sodium chlorate(I) solution (NaClO). This is produced when chlorine is reacted with sodium hydroxide solution:

$$Cl_2(g) + 2NaOH(aq) \rightarrow NaCl(aq) + NaClO(aq) + H_2O(l)$$

Domestic bleaches contain sodium chlorate(I) solution. The chlorate(I) decomposes to give oxygen which acts as an **oxidizing agent**:

NaClO + coloured material → NaCl + oxidized
(decolorized)
material

boiling   The change from the liquid state to the gas state at a particular temperature, the **boiling point**. Boiling takes place when the **vapour** pressure of the liquid equals the pressure of the atmosphere above the liquid.

boiling point The temperature at which **boiling** occurs. It is not fixed, but instead depends on the atmospheric pressure. The higher the pressure, the higher the boiling point. Boiling points are usually quoted at one **atmosphere**. Boiling temperatures are also affected by impurities in the liquid: the presence of impurities causes the boiling temperature to rise. *See also* **evaporation**.

bonds The strong forces that hold together atoms in molecules and **giant structures**. The forces exist between the atoms and are generated by electrons. Either electrons are shared between atoms, or atoms gain or lose electrons forming **ions**.
  The sharing of electrons (*see* **covalent bonds**) usually occurs when two non-metallic atoms join together, e.g. H–H or H–Cl. The loss or gain of electrons (**ionic** or electrovalent bonding) occurs when atoms of metallic elements join with those of non-metallic elements, e.g. Na + Cl or Mg + O.
  See also **metallic bond**.

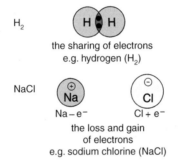

bonds *Covalent bonding in the hydrogen molecule and ionic bonding in sodium chloride.*

bond energy value The amount of energy needed to break a chemical **bond**.

boron A non-metallic element in group III of the **periodic table**. Its **oxide** is used to make glass (*see* **borosilicate glass**). It is a component of **alloys** and is a moderator in nuclear reactors.

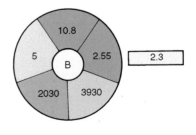

boron

borosilicate glass **Glass** made by the addition of boron oxide (B_2O_3) during its manufacture. This results in a material that does not expand very much.

Consequently the glass can be heated or cooled rapidly without cracking. 'Pyrex' glassware contains borosilicates.

Boyle's law   A law that states that at constant temperature the volume of a fixed **mass** of gas is inversely proportional to the pressure of the gas. In other words, if the pressure is doubled the volume is halved. *See also* **Charles' law, gas laws**.

$$P \propto \frac{1}{V}$$

$$PV = \text{constant}$$

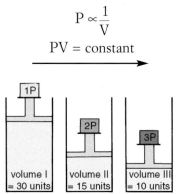

Boyle's law *As the pressure doubles the volume is halved.*

brass   An **alloy** of copper and zinc which usually contains about 30% zinc. Brass is an attractive yellow-golden coloured material which is used for ornaments and for electrical components, e.g. in electrical plugs.

breath test   A test used to measure the amount of **ethanol** drunk in alcoholic drinks by a person. A person blows into a machine which analyses the amount of ethanol in the breath. The machine converts this reading into a measure of the amount of ethanol in the blood. The current legal limit is 80mg of ethanol per 100ml of blood. Above this limit a driver is committing an offence in the UK.

brewing   A process which involves the **fermentation** of sugars to the alcohol **ethanol**. Beers, wines and spirits are produced by the brewing and distilling industries.

brine   A solution of sodium chloride in water. The solution is more concentrated than seawater and is used in the food-processing industry, e.g. in the production of bacon and the preservation of many foods. In the chemical industry chlorine is produced by the **electrolysis** of brine. *See also* **curing**.

brittle   Easily snapped.

bromide A compound of bromine and another element. Metal bromides are usually **ionic** solids, e.g. sodium bromide NaBr. Non-metal bromides are **covalent compounds**, e.g. hydrogen bromide HBr. *See also* **halides**.

bromine (Br) A member of the **halogen** group of elements. It is a **volatile** red liquid at room temperature. The liquid is very corrosive and the vapour is both irritating and poisonous.

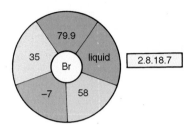

bromine

    Although it is less reactive than chlorine, bromine will react vigorously with metals forming **bromides**:

$$Mg(s) + Br_2(l) \rightarrow MgBr_2(aq)$$

Bromine is extracted from seawater by treatment with chlorine and sulphur dioxide. One thousand litres of seawater contains about 65g of bromine.

% of production	Use
24	Fuel additives
20	Flame retardants
18	Oil well fluids
13	Agrochemicals
6	Dyes
4	Water purification
15	Other

bromine *The uses of bromine.*

bronze The combination of copper (>90%) with tin (<10%) which results in an **alloy** that is much stronger than copper. The discovery of bronze in the Middle East (3000 BC) gave rise to important changes in the way humans lived (the Bronze Age). Nowadays, bronze is mainly used for gear wheels and engine bearings.

Brownian motion The random movement of particles suspended in a liquid or gas, such as pollen in water or smoke in air. The explanation is that the microscopic particles which make up the liquid or gas, e.g. water and air molecules, are moving randomly and are hitting against the larger particles of pollen or smoke, causing them to move. Brownian motion is put forward as evidence for the **kinetic theory**.

buckyballs *See* **carbon**.

burette A long glass tube with a tap on the end. The tube is marked (graduated) every 1cm³ with smaller graduations of 0.1 cm³. It is used in **titrations** to accurately measure small volumes (up to 50 cm³) of liquid. The volume under the last graduation is called the *dead space* because it is not used for measurements. *See also* **pipette**.

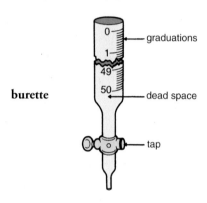

burner or Bunsen burner A device invented in the 19th century so that coal-gas could be burnt cleanly giving a hot flame. **Natural gas** is now used. With the air hole open the gas is mixed with air and this makes sure that all the gas is burnt and no soot is produced. With the air hole shut a yellow sooty flame is produced because the gas is not fully burnt. *See also* **flame**.

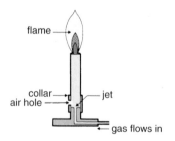

burner *The components of a Bunsen burner.*

burning *See* **combustion**.

butane An **alkane** compound. It is a gas at room temperature but it is easily liquefied. It is present (dissolved) in **petroleum** and is used in refineries both as a fuel and as a starting material for the production of hydrogen and petrol components. *See* **refining**.

Butane is a major component of **liquefied petroleum gas**. The gas burns in air:

$$2C_4H_{10}(g) + 13O_2(g) \rightarrow 8CO_2 + 10H_2O(g)$$

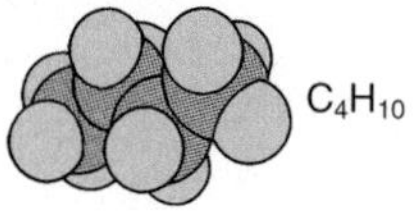

C_4H_{10}

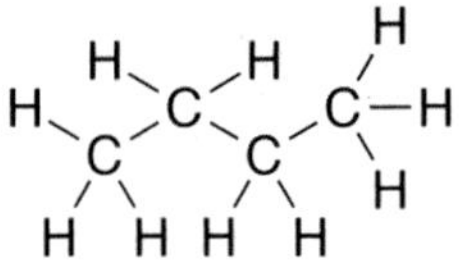

butane *An alkane compound.*

by-product   Something which is produced in a reaction in addition to the product which is required. For a manufacturing process to be economic, it is important that by-products are sold.

C

calcite A mineral form of calcium carbonate ($CaCO_3$) which is found in **limestone** and **chalk**.

calcium A soft, metallic element that is in group II of the **periodic table**. It is a fairly reactive metal, giving a slow but steady stream of hydrogen when it is added to cold water:

$$Ca(s) + 2H_2O(l) \rightarrow Ca(OH)_2(aq) + H_2(g)$$

3.6% of the Earth is made up of calcium compounds. Many of these are very common, e.g. **limestone** and **chalk** ($CaCO_3$). Calcium is obtained by the **electrolysis** of molten calcium chloride.

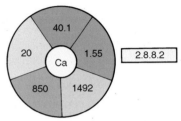

calcium

calcium compounds

Calcium carbonate $CaCO_3$	Used as building stone and in the production of cement. Lime (calcium oxide) is produced from calcium carbonate in a lime kiln.
Calcium chloride $CaCl_2$	Used to produce calcium metal. The anhydrous salt is used as a drying agent.
Calcium hydrogencarbonate $Ca(HCO_3)_2$	An important cause of temporary hardness in water.
Calcium hydroxide (slaked lime) $Ca(OH)_2$	A slightly soluble alkali. The aqueous solution is called limewater. It is used to make mortar.
Calcium oxide (lime) (CaO)	Important uses in agriculture to combat excess acidity in soil.
Calcium phosphate $Ca_2(PO_4)_2$	An essential part of bones and teeth.
Calcium sulphate $CaSO_4$	Gives rise to permanent hardness in water. It is found as anhydrite and gypsum.

cane sugar *See* **sucrose**.

carbohydrates **Organic** compounds that contain the elements carbon, hydrogen and oxygen only. They are made by green plants through **photosynthesis**. Their formulae are always of the form: $C_x(H_2O)_y$, e.g. **sugars** such as **glucose** ($C_6H_{12}O_6$), **sucrose** ($C_{12}H_{22}O_{11}$) and ribose ($C_5H_{10}O_5$) and **polymers** such as **starch** ($C_6H_{10}O_5$)n and **cellulose** ($C_6H_{10}O_5$)$_n$. *See also* **monosaccharide, disaccharide, polysaccharide**.

carbon   A non-metallic element which is in group IV of the **periodic table**. It occurs in nature as two crystalline **allotropes**: **diamond** and **graphite**. Further crystalline allotropes, first identified in 1985, were found which have a cage structure. These are known as *fullerenes* or *buckyballs* and have the formulae C_{50}, C_{60} and C_{70}. Carbon also exists in **amorphous** forms.

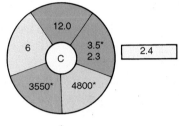

carbon

All living tissue contains carbon compounds (**organic** compounds), e.g. **carbohydrates, proteins, fats**, and without these compounds life would be impossible.

Although the element burns in oxygen or air it is otherwise fairly unreactive. It is an important **reducing agent. Coke, coal** and **charcoal** are all impure forms of carbon. Carbon is also found combined in materials such as **limestone**, in the form of metal carbonates.

carbonate   Compounds containing the $-CO_3^{2-}$ ion, which has a **valency** of 2. Metal carbonates occur widely in nature, e.g. **limestone** ($CaCO_3$), dolomite ($CaMg(CO_3)_2$) and **malachite** ($CuCO_3.Cu(OH)2$). Calcium carbonate plays a part in the **carbon cycle**.

Carbonates of all metals (except group I metals) are insoluble in water. All carbonates produce carbon dioxide when heated strongly (K_2CO_3, Na_2CO_3 with difficulty) and when treated with dilute acid.

The chemical test of a carbonate is to add acid to the solid compound and to pass the gas produced through **limewater**. A **precipitate**, seen as a milkiness, indicates that the solid was a carbonate.

carbon compounds
Carbon dioxide CO_2
   This gas can be formed by the action of heat or acids on carbonates or by the complete combustion of carbon. It is also produced in the **fermentation** of sugars. The test for carbon dioxide is that it turns limewater milky. Carbon dioxide plays a vital role in photosynthesis and the carbon cycle. *See also* **greenhouse gas**.

Carbon monoxide CO
This gas is formed when carbon or its compounds are not completely burnt. It is an air pollutant, being produced ininternal combustion engines and from the burning ofcigarettes. It is a very poisonous gas because it combineswith the haemoglobin in blood. It is useful as a reducingagent in the blast furnace. The gas burns to give carbon dioxide.

See also **carbonates, carbohydrates, hydrocarbons.**

carbon cycle   The process in which carbon (mostly in carbon dioxide) is circulated around the Earth. As plants and animals respire, and **fossil fuels** are burnt, carbon dioxide is released into the atmosphere. At the same time, it is continually taken out of the atmosphere by the process of **photosynthesis**, to build up plant structures. The plants are eaten by other organisms which breathe out the carbon dioxide – and so the cycle continues.

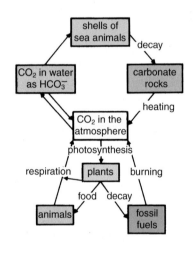

There are stores of carbon in the oceans, lakes and rivers (as dissolved CO_2) and in **carbonate** rocks and fossil fuels. A simple diagram of these movements is shown here. Some changes occur quickly, e.g. burning; others very slowly, e.g. formation of rocks.

carbon cycle

carbon dating   A method of telling the age of once-living material. Carbon has a radioactive **isotope** ($^{14}_{6}C$) and the atmospheric carbon dioxide contains a small constant proportion of this isotope. All carbon compounds in living tissue contain this carbon-14 isotope in the same constant proportion. Plants take in the isotope through **photosynthesis** and animals take in the isotope by eating plants. When an animal or plant dies, however, the carbon-14 is not replenished and the proportion falls as the isotope decays. The **half-life** of $^{14}_{6}C$ is known to be 5730 years. By measuring the amount of radioactivity in a dead material it is therefore possible to estimate the age of the material. The technique is used by archaeologists.

carboxylic acids   Organic **acids** with the formula R–COOH. *See* **ethanoic acid**.

carotene (ß-carotene)   Natural yellow colouring. It is found in carrots, egg yolk and butter and in all green leaves. It is now widely used as an additive in foods to provide yellow/orange colouring.

cast iron   *See* **blast furnace**.

catalysis   *See* **catalyst**.

catalyst   A substance that alters the **rate** of a chemical **reaction**. The catalyst remains unchanged at the end of the reaction. The process is called *catalysis*. **Transition metals** and their compounds are often useful catalysts. Some examples are shown below:

Reaction	Catalyst
Haber process	Iron
Contact process	Vanadium(V) oxide (V_2O_5)
Ammonia→nitric acid	Platinum/rhodium alloy
Hardening fats	Nickel

catalyst *Transition metal catalysts.*

The chemical processes that go on inside animals and plants are nearly all dependent on organic catalysts called **enzymes**.

catalytic convertor   Part of the exhaust system of modern petrol engines. A platinum/rhodium **catalyst** in a honeycomb structure converts carbon monoxide (CO), nitric oxide (NO) and unburnt **hydrocarbon** compounds from the petrol into carbon dioxide (CO_2), nitrogen (N_2) and nitrous oxide (N_2O). This is one way of reducing the amount of air pollution caused by the internal combustion engine.

Unleaded petrol must be used when a car is fitted with a catalyser otherwise the catalyst will be 'poisoned' by the lead and rendered useless. *See also* **greenhouse effect**, **greenhouse gases**, **ozone layer**.

catalytic cracking   *See* **cracking**.

catalytic hydrogenation   *See* **hydrogenation**.

cathode   The **electrode** that carries the negative charge in a solution that is undergoing **electrolysis**. Positively charged **ions** (**cations**) are attracted to the cathode. They gain electrons at the cathode. *Compare* **anode**. *See* figure on next page.

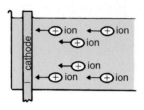

cathode *Positive ions attracted to the negative cathode.*

cation   A positively charged **ion**. Most cations are metal ions, e.g. Fe^{2+}, Cr^{3+}, Na^+, Ca^{2+}, Al^{3+} and are attracted to the **cathode** during **electrolysis**. Three cations not formed from metals are the **hydrogen ion** (H^+), the **oxonium ion** (H_3O^+) and the **ammonium ion** (NH_4^+).

caustic   Capable of burning or corroding organic material, e.g. flesh. The word is usually restricted to use with **alkaline** materials, e.g. caustic soda (sodium hydroxide) and caustic potash (potassium hydroxide).

cell   1. A device that converts chemical energy into electrical energy. Chemical reactions occur that cause electrons to flow through a circuit. Two or more cells joined together is called a **battery**.
2. Apparatus and chemicals which are used in **electrolysis**: an *electrolytic cell.*

cellulose   A carbohydrate **polymer** made up of **glucose monomers**; the formula is $(C_6H_{10}O_5)_n$. Cellulose is the material from which cell walls of plants are made. Cellulose is used to manufacture paper and **rayon**.

Celsius scale of temperature (°C)   A scale of **temperature** based on a 100-degree range between the melting point of pure ice (0°C) and the boiling point of pure water at a pressure of one **atmosphere** (100°C). It was originally called the *centigrade* scale. One degree Celsius equals one **kelvin**.

cement   A substance made by heating chalk or limestone together with clay, and then powdering the product. The result is a mixture of calcium silicate and calcium aluminate, which is usually a grey colour. When water is added a corrosive mixture is produced that is **alkaline**. Cement is a very important **adhesive** material in the building industry:

cement + sand + water $\rightarrow$ mortar

cement + sand + gravel + water $\rightarrow$ concrete

centigrade   *See* **Celsius scale of temperature**.

ceramics   **Inorganic**, crystalline materials which usually have the following characteristics:
(a) great hardness and resistance to wear;
(b) very high melting point;
(c) they are chemically **inert**;
(d) they are poor **conductors** of heat and electricity;
(e) they are **brittle**, and not **malleable**, **ductile** or **plastic**;
(f) they are opaque.
Ceramic materials are mostly compounds of metallic and non-metallic elements such as silicon, aluminium, carbon, nitrogen, oxygen. They are more rigid and inflexible than metals. This makes them harder than metals, but usually also more brittle. Ceramics include **glass**, **enamel**, pottery, **clay** products, **abrasive** and **refractory** material.

CFCs (chlorofluorocarbons)   **Organic** compounds which contain carbon, chlorine and fluorine atoms. CFC molecules were developed in the 1920s for use as refrigerants. They are stable, **inert**, odourless, non-toxic, non-corrosive and do not burn. After World War II, their use grew rapidly, for example, as propellants in **aerosol** containers, fire-extinguishing fluids, in (synthetic) **foam** production, and as fluids in air-conditioning systems and freezers.

In the mid-1970s, it became known that CFCs break down in the stratosphere and produce reactive chlorine atoms which then go on to react with **ozone** molecules. In this way, the amount of ozone in the stratosphere is reduced and we have less protection from the harmful effects of **ultraviolet radiation**. Now efforts are being made to reduce the use of CFCs, and to find substitutes for them which do not reduce the amount of ozone in the stratosphere. Unfortunately, some of the readily available alternatives are **greenhouse gases**. *See also* **Earth's atmosphere**.

chain   A line of atoms of the same type in a molecule. Carbon atoms form chains easily, and in **polymers**, these straight chains can be thousands of atoms long. The chains can have branches too.

chain reaction   A rapid series of **reactions** in which the product of each reaction causes the next one to occur. For example, some **isotopes** are unstable and when they are bombarded with a **neutron** they break up to produce smaller atoms and more neutrons (nuclear **fission**). An example is the bombardment of uranium-235:

$$^{235}_{92}U + \text{neutron} \rightarrow \left\{ \begin{array}{l} \textbf{atoms of barium} \\ \text{and } \textbf{krypton} \end{array} \right\} + 3 \text{ neutrons}$$

In the example, each time a neutron is used up, three more are produced, each of which can then go on to split up another large unstable atom and produce even more neutrons. As the reaction goes on, more and more neutrons are produced. This is a chain reaction and is the basis for **nuclear reactions** in nuclear power stations and atom bombs.

chalk   A type of rock formed from the shells of marine animals. It is mainly calcium carbonate ($CaCO_3$). Chalk is a softer rock than **limestone** and its uses are limited by this.

change of state **or** phase change   The movement of a material to and from the **solid**, **liquid** and **gas** states. Such movements are always accompanied by energy changes. *See also* **kinetic theory**.

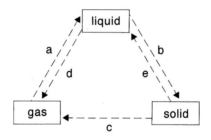

change of state *(a) condensation, (b) freezing, (c) sublimation, (d) vaporization, (e) melting.*

charcoal   A black, impure form of carbon made by heating organic material (usually bones or wood) in a limited supply of air. In this process, **volatile** material escapes and the resulting charcoal is composed mainly of carbon. Before coal was widely available, charcoal was used to make iron from iron ore.

A particular form of charcoal, *active carbon*, which is specially made, is able to adsorb small molecules onto its surface. This is used in gas masks to absorb poisonous gases; and in various industrial processes where coloured impurities need to be removed from materials, such as sugar refining to produce white sugar.

Charles' law   The volume of a fixed mass of gas is dependent upon its temperature. If the temperature of a gas is doubled (measured on the **Kelvin** scale) then the volume of the gas will double provided that the pressure of the gas is kept constant. *See also* **Boyle's law**, **gas laws**.

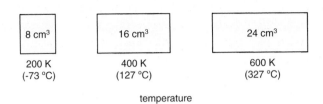

Charles' law *Doubling the temperature of a gas doubles its volume.*

chemical change   A change in which one or more chemical substances are changed into different ones. Such a change is usually accompanied by the giving out or taking in of **heat energy**. *See also* **physical change**.

chemical energy   *See* **energy**.

chemical properties   *See* **properties**.

chemotherapy   The use of chemicals to treat illness and pain.

chlor-alkali industry   The production of hydrogen, sodium hydroxide and chlorine by the **electrolysis** of sodium chloride solution.

chlorides   Compounds of chlorine and another element. Metal chlorides are usually **ionic** solids, e.g. sodium chloride (NaCl) and barium chloride ($BaCl_2$). Non-metal chlorides are **covalent compounds** and are usually either low boiling point liquids such as tetrachloromethane (CCl_4) or gases such as hydrogen chloride (HCl).
   Metal chlorides react with concentrated acids to produce hydrogen chloride gas:

$$H_2SO_4(l) + NaCl(s) \rightarrow NaHSO_4(s) + HCl(g)$$

The test for chlorides is to mix a solution with silver nitrate solution. If a white **precipitate** forms which dissolves when ammonia solution is added, the substance is a chloride.

chlorination   **1.** The addition of chlorine to drinking water and to water used in swimming pools, in order to kill dangerous bacteria.
   2. Reactions between chlorine and **hydrocarbons** that produce chlorinated hydrocarbons.

chlorine (Cl)   A green gas at room temperature. It is a member of the **halogen** group of elements (group VII) and is very reactive. It has a choking effect and attacks lung tissue and the throat if it is breathed in. It was used as a chemical weapon in World War I.

Chlorine occurs naturally as **chlorides** and sodium chloride is abundant in seawater. Chlorine is extracted by the **electrolysis** of **rock salt** solutions. It is used widely to make **polymers, biocides, disinfectants** and **solvents**. A solution of chlorine in water is used as a **bleach**. *See also* **chlorination**.

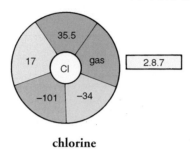

chlorine

Chlorine is a vigorous **oxidizing agent** which readily reacts with most elements. It is made in the laboratory by the oxidation of concentrated hydrochloric acid:

$$4HCl(aq) + MnO_2(s) \rightarrow Cl_2(g) + MnCl_2(aq) + 2H_2O(l)$$

chloroethene **or** vinyl chloride   The **monomer** from which **poly(chloroethene)** (PVC) is made, by reacting ethene (C_2H_4) with chlorine (Cl_2).

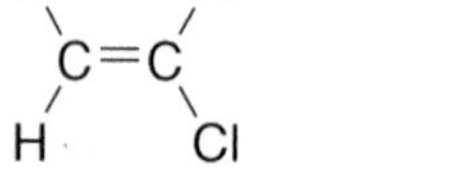

C₂H₃Cl

chloroethene
(a) *The monomer.*

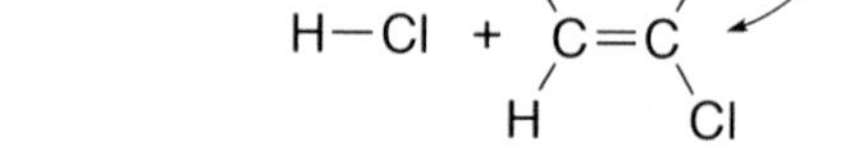

chloroethene (b) *Reacting ethene with chlorine to make chloroethene.*

chlorofluorocarbons   *See* **CFCs**.

chlorophyll   The green pigment in plants that is responsible for the absorption of the Sun's energy by the plant. This energy is used in the process of **photosynthesis**. There are two kinds of chlorophyll, each of which is a complex molecule that contains a magnesium atom.

chromatography A technique for separating **mixtures** of **solutes** in a **solution**. The process occurs because some materials move through paper at different rates: in the example, dye A moves through the filter paper more quickly than dye B. The technique is widely used in **analysis**.

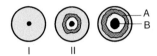

chromatography *A spot of dye is placed on filter paper (i); as the water moves out across the paper the dye separates (ii) in different bands of colour.*

chromium A **transition metal**. It has important uses in **alloys** such as **stainless steel** and is used in chromium plating on items such as bicycle handlebars, kettles, and cutlery. It is also used in the production of **pigments**, wood preservatives and in the tanning industry. It has a high resistance to corrosion. *See also* **dichromate(VI) ion**, **electroplating**.

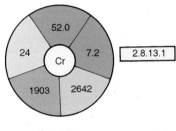

chromium

citric acid A weak organic **acid** that occurs naturally in citrus fruits such as oranges and lemons.

clay A natural material containing aluminium, silicon and oxygen atoms (as various kinds of aluminium silicate) which is commonly found in soils. Clays are used in pottery making, in **ceramics** and as fillers in the manufacture of rubber, paint, plastics and paper. Clay has an important use as a lubricating mud during drilling for minerals.

cm³ (cubic centimetre) A unit of **volume** used in scientific work. It is the volume of a cube which has sides of 1 cm. 1000 cm³ are equal to 1 litre or 1 cubic decimetre (1 **dm³**).

coal A **fossil fuel** consisting of fossilized plant material which was living millions of years ago. It has a very complex chemical structure containing compounds made up of carbon, hydrogen, oxygen, nitrogen and sulphur. Coal is used as a fuel in power stations, industry and the home and was, before the use of **natural gas**, used as the source of coal gas. About 20% of coal is used to make **coke**.

If coal is heated in the absence of air, coal tar is produced. In the middle of the 20th century coal tar was an important source of organic chemicals, e.g. **phenol** from which dyes, drugs and **polymers** were made. Now over 90% of such products come from **petroleum** sources.

As the world has greater reserves of coal than petroleum, attempts are being made to convert coal into petroleum products.

cobalt   A **transition metal**. It is a magnetic element and is used alloyed with iron in the manufacture of magnets. The radioactive isotope $^{60}_{27}\mathrm{Co}$ emits **gamma rays** and it is used in the treatment of cancers. Cobalt(II) chloride is blue when **anhydrous** and pink when **hydrated**, and so the compound is used as a test for the presence of water.

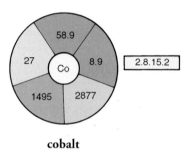

cobalt

coke   The material left behind when the **volatile** compounds have been removed from **coal**. It contains over 80% carbon. It was formerly made during the production of coal gas but is now made in *coke ovens* from special *coking coal*. Coke is used to make carbon monoxide in the **blast furnace** and is an important (smokeless) fuel.

colloidal dispersion or colloid   A mixture in which the **particles** are smaller than those found in **suspensions** but larger than those found in **solutions**. Particles in colloidal dispersions are between 1 nm and 1000 nm long (*see* **nanometre**). Colloidal dispersions cannot be separated by a **filter**. The following are examples of various types of colloidal dispersion: **aerosol** (liquid/solid dispersed within a gas); **foam** (gas within a solid or liquid), **emulsion** (liquid within a liquid). *See also* **electrophoresis, gel**.

combining power   *See* **valency**.

combustion   A burning reaction in which a substance combines with a gas. Heat and light (i.e. **flame**) usually accompany combustion reactions. Most combustions involve oxygen, e.g.:

$$2H_2(g) + O_2(g) \rightarrow 2H_2O(g)$$

common salt   *See* **sodium compounds**.

complex ion   A **cation** (e.g. a metal **ion** or a hydrogen ion) which is bonded to one or more small molecules by a **coordinate bond**. The simplest examples are the **ammonium ion** and the **oxonium ion**, where the **lone pair of electrons** on the nitrogen and oxygen atoms forms the bond.

Transition metal ions form a large number of complex ions because they can easily accept the donated electron pairs.

NH_4^+ H_3O^+

complex ions *Ammonium and oxonium ions.*

composite material   A material made up of two or more components whose properties combine to give something more useful than the separate components. Composites are usually made up of layers of materials reinforced by **fibres** (for example of carbon or glass). Composite materials are strong and are widely used in cars, planes, boats, furniture and sports gear. *See also* **fibre-reinforced plastics**.

compost   A natural **fertilizer**, made by the decay of plant materials by bacteria.

compound   A pure substance which is made up of two or more **elements** chemically bonded together. The **properties** of a compound are quite different from the properties of the elements bonded together within it. Compounds may contain ionic or covalent bonding (*see* **ionic bonds**, **covalent bonds**). Examples are methane CH_4, water H_2O, sodium chloride $NaCl$. *Compare* **mixture**.

concentrated   Containing a high proportion of something. A concentrated **solution** is one which contains a relatively large proportion of **solute**. If you want to increase the concentration of a solution, this can be done by either adding more solute or by removing **solvent** (e.g. by **distillation**) from the solution. *Compare* **dilute**.

concentration   A measure of how much **solute** is dissolved in a **solution**. It is usually expressed in terms of how much substance is present in a given volume of the solution. This can either be in terms of **mass**, e.g. grams, or in terms of how many particles, e.g. **moles**. Examples of concentrations and how they can be written are, for a solution containing 170 grams of silver nitrate (1 mole) in one cubic decimetre (**dm^3**) of solution:

$$170 \text{ g dm}^{-3}, \text{ or } 1 \text{ mol dm}^{-3} \text{ or } 1 \text{ mol l}^{-1} \text{ or } 1 \text{ M}$$

concrete   A building material made up of an **aggregate** (crushed stone, or **slag**) mixed with **cement**, **sand** and water.

condensation **1.** The change from the gas or vapour state to the liquid state, e.g. water condensing onto a cold window pane.
2. A chemical reaction in which two or more molecules react to produce a larger molecule and a small molecule such as water. This is one method of producing **polymers**, such as **nylon**.

condensation polymer A **polymer** made by a **condensation** reaction.

conductor A substance that allows **heat energy** or electricity to flow through it. An *electrical conductor* is a substance which will allow an electric **current** to flow through it. Metals, solutions that contain **ions**, and molten **ionic** compounds are electrical conductors; all other substances are **insulators**. Graphite is a notable exception. Certain **metalloids** are **semiconductors**. Metals are also good conductors of heat.

conductivity A measure of the extent to which a **conductor** allows an electric current to flow.

conductivity titration A **titration** which can be followed by measuring the changes in the **conductivity** of a solution.

conservation of energy and mass (law of) The law that states that **mass** and **energy** cannot be created or destroyed in a chemical **reaction** because the mass of the reacting substances and the mass of the products are equal. They are thus said to be *conserved*. Similarly:

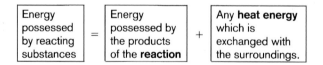

These relationships are only approximations. The work of Einstein showed that mass and energy can be in fact converted into each other. This is important in **nuclear reactions** but in ordinary chemical reactions the relationship holds good.

contact process
The method used to make almost all the **sulphuric acid** produced today. The flow chart below shows the steps. Most sulphuric acid is now made from sulphur. The reactions marked * are **exothermic** and so the process is relatively cheap to run because little **heat energy** has to be supplied.

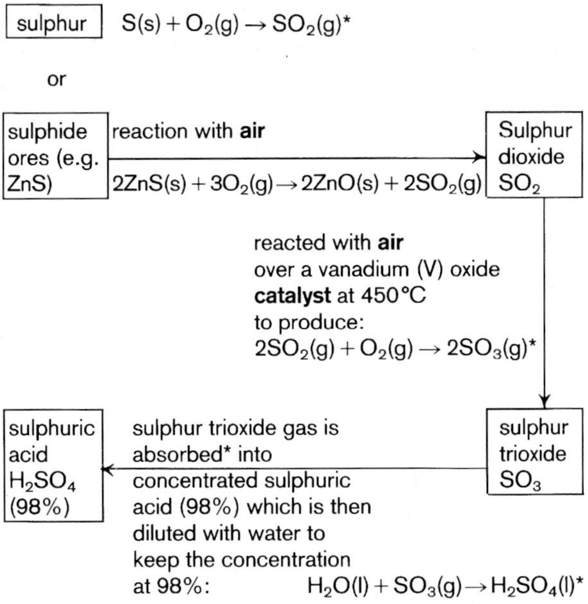

sulphur	$S(s) + O_2(g) \rightarrow SO_2(g)*$

or

sulphide ores (e.g. ZnS)	reaction with **air**	Sulphur dioxide
	$2ZnS(s) + 3O_2(g) \rightarrow 2ZnO(s) + 2SO_2(g)$	SO_2

reacted with **air** over a vanadium (V) oxide **catalyst** at 450°C to produce: $2SO_2(g) + O_2(g) \rightarrow 2SO_3(g)*$

sulphuric acid H_2SO_4 (98%)	sulphur trioxide gas is absorbed* into concentrated sulphuric acid (98%) which is then diluted with water to keep the concentration at 98%: $H_2O(l) + SO_3(g) \rightarrow H_2SO_4(l)*$	sulphur trioxide SO_3

contact process *The steps in making sulphuric acid.*

control   A parallel experiment, carried out at the same time as a main experiment, in which the factor being investigated is kept constant. The result of the main experiment in which this factor is varied can be compared with the control to check the extent of any change.

coordinate bond or **dative bond**   A **covalent bond** in which both the electrons in the bond have come from the same atom. Electrons have been donated by one atom only, rather than one electron being provided by each atom. In coordinate bonding, the donor atom uses a **lone pair of electrons**. *See also* **complex ion**.

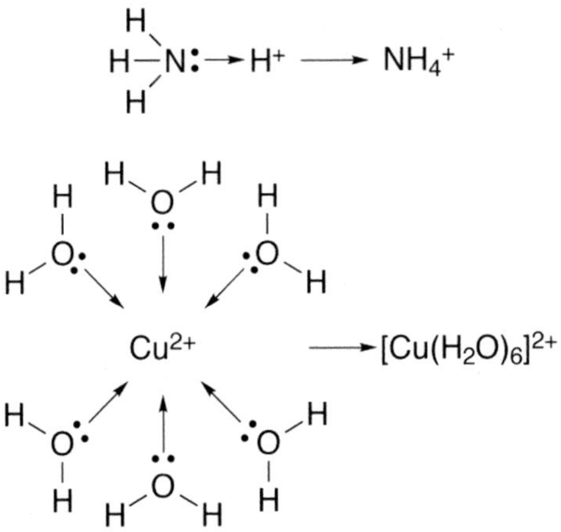

$H-\overset{\displaystyle H}{\underset{\displaystyle H}{N}}\!:\; \longrightarrow H^+ \longrightarrow NH_4^+$

$Cu^{2+} \longrightarrow [Cu(H_2O)_6]^{2+}$

coordinate bond *Examples. The large arrows represent coordinate bonds.*

copolymer   A **polymer** made by the joint **polymerization** of two (or more) kinds of **monomer**. Copolymers are **compounds** rather than **mixtures** and can be produced with particular properties in mind. They are widely used in industry to produce everyday objects. The fibre Acrilan is a good example.

copper   A **transition metal** that plays an important part in our lives. It is a vital trace element in our bodies – we need between one and two milligrams per day.

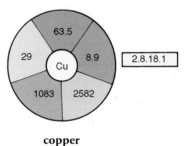

copper

Copper is an unreactive metal, coming low down in the **electrochemical series**. It will not liberate hydrogen from dilute acids.

Copper can be purified by using the impure metal as the **anode** with copper(II) sulphate as the **electrolyte** and a small pure copper cathode in an **electrolysis** cell.

Copper metal is used in plumbing, for coinage and for electrical wiring (copper is an excellent **conductor**). It is extracted commercially from its **sulphide** ores, chalcopyrite ($CuFeS_2$) and bornite (Cu_5FeS_4) and its carbonate ores, malachite ($CuCO_3.Cu(OH)_2$) and azurite ($2CuCO_3.Cu(OH)_2$).

Copper is also found as the free element, an indication of its lack of **reactivity**.

copper compounds

Copper(II) chloride $CuCl_2.2H_2O$	Used to give fireworks a green colour and in the removal of sulphur from petroleum.
Copper(II) hydroxide $Cu(OH)_2$	Used in the manufacture of rayon, in dyeing textiles and as a blue pigment.
Copper(II) carbonate $CuCO_3$	Forms part of the ores malachite and azurite – both basic copper carbonates. Another basic carbonate – verdigris – is formed as a surface layer on copper which is exposed to the atmosphere.
Copper(II) copper sulphate	Forms blue crystals and can be made by treating compounds (oxide, carbonate, hydroxide) with dilute

$CuSO_4.5H_2O$ sulphuric acid. The anhydrous salt is a white powder
that turns blue when water is added:

$$CuSO_4 + 5H_2O \rightarrow CuSO_4.5H_2O$$
white   blue

This reaction is often used as a test for the presence of
water. The compound is used in wood preservatives
and fungicides, e.g. Bordeux mixture.

corrosion A process whereby stone or metal is chemically eaten away.
Good examples are the weathering of limestone buildings by rainwater,
which contain dissolved acids, and the rusting (*see* **rust**) of steel. Corrosion
begins at the surface and often a surface layer is formed that protects the
rest of the material, e.g. the **oxide** coating on aluminium. A protective layer
is not formed in rusting; the rusting process goes through the steel until it
is all corroded. That is why rust is so damaging and causes great expense.
Corrosion can be prevented by means of **sacrificial anodes, paint** and
electroplating. *See also* **acid rain**.

cotton A natural **cellulose** fibre which forms strong and hard-wearing
fabrics. Cotton is good to wear next to the skin in hot climates because of
its ability to absorb water.

coulomb A measurement of electric charge. It is the product of the
current (measured in **amperes**) and the time (measured in **seconds**). For
example, if a current of 0.6 amps flows for 10 seconds, 6 coulombs of
electricity will have passed through the circuit.

covalent bonds Chemical **bonds** formed by two atoms coming together
and sharing their electrons. For example, hydrogen atoms have one
electron each (denoted by • in the diagram below). The atoms form a
molecule of hydrogen (H_2) by the two electrons forming a bond. Thus each
'shared pair' of electrons produces one covalent bond.

hyrogen atoms (H)

hydrogen molecule (H_2)

covalent bonds *Atoms of hydrogen form a molecule*
by two electrons forming a bond.

Covalent bonds are usually found in compounds which only contain non-
metallic elements, e.g. hydrogen chloride (HCl).

They are also found in elements with an
atomicity of two or more, for example
oxygen (O_2) and hydrogen (H_2).

hydrogen chloride

   Covalent bonds that contain two
electrons, i.e. one 'shared pair' are termed
single bonds. Many molecules contain
double bonds or **triple bonds**. Covalent

covalent bonds *Hydrogen chloride.*

bonds within molecules are not as strong as **ionic bonds**. Covalent bonds
are also found within some **giant structures**, e.g. silicon dioxide (SiO_2) and
silicon carbide (SiC). Here, the bonding is strong and these compounds
tend to have high melting points. *See also* **covalent compound**.

covalent compound   A **compound** that contains **covalent bonds**. Such
compounds tend to be non-**conductors**. Covalent compounds have low
melting points and boiling points when they have molecular structures;
examples are carbon dioxide, water, nitrogen. Giant structures that contain
covalent bonds, however, have high melting and boiling points. Examples
are silicon dioxide, boron nitride.

cracking   The process in which large hydrocarbon molecules are broken
up into smaller molecules. This process is the basis of the **petroleum**-refining
industry. Small molecules are more valuable than larger ones because they
are the starting materials needed in the production of other chemicals such
as **polymers**, e.g. **poly(ethene)** and **poly(propene)**, and **fuels**, such as **petrol**
and **diesel fuel**.

   Different products are formed from different raw materials under
different conditions, as shown in the table.

Conditions	Process name	Products
High temperature	Thermal cracking	Unsaturated and saturated molecules are produced
High temperature in the presence of steam	Steam cracking	
High temperature in the presence of a catalyst	Catalytic cracking	
High temperature and pressure in the presence of hydrogen	Hydro-cracking	The product is totally saturated

cracking *Different processes and their products.*

critical temperature *See* **vapour**.

cross-linking The formation of chemical **bonds** between **polymer** molecules lying side by side. Cross-linking within molecules produces materials that
(a) have higher **relative molecular masses**;
(b) are tougher and less flexible;
(c) have higher melting points;
(d) are less soluble.
Examples of materials that are extensively cross-linked include rubber which has undergone **vulcanization**, bakelite and similar **thermosetting polymers**, and **epoxy resins**.

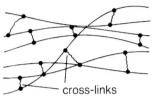

cross-links

cross-linking *Chemical bonds formed between polymer molecules.*

crude oil *See* **petroleum**.

cryolite A mineral form of sodium aluminofluoride, Na_3AlF_6. In the extraction of aluminium, **bauxite** (Al_2O_3) is dissolved in molten cryolite at 900 °C and the mixture is then electrolysed. Aluminium is produced at the **cathode**. Cryolite only occurs naturally in any quantity in Greenland. The amounts needed in the extraction of aluminium have to be produced by the reaction between aluminium oxide (Al_2O_3) and hydrogen fluoride (HF).

crystal A solid substance in which the atoms (molecules or ions) are arranged in a set geometrical three-dimensional pattern (*see diagram*). The shape of these atomic arrangements (crystal **lattices**) determines the shapes that the solid has when it undergoes **crystallization**.

cubic octahedral tetrahedral

crystal *Examples of different crystal arrangements.*

crystallization The process in which **crystals** are formed. In nature, crystals are produced when molten rocks cool down and solidify. In laboratory experiments, crystals are usually produced from a solution. There are two main methods:
(a) a solution is left at room temperature. Slow evaporation of the **solvent** takes place and crystals are left behind;

(b) a **supersaturated solution** is made above room temperature. As this is cooled down, crystals are produced.

curing　**1.** The hardening of resins.
　2. A method of preserving meats such as bacon, ham, gammon and beef. The curing solution usually consists of a mixture of sodium salts and is either rubbed onto or injected into the raw meat.

current　The movement of **electrons** through a **conductor**. It is measured in **amperes** (amps). By convention, electric current is said to move from the positive terminal to the negative terminal. In reality, however, electrons travel from the negative terminal to the positive. This difference is because the electron carries a negative electric charge.

cyanides　A range of compounds that contain the –CN group of atoms. Potassium cyanide (K^+CN^-) and hydrogen cyanide ($H–C{\equiv}N$) are the best known examples. They are very poisonous.

cyclic compounds　Carbon compounds in which atoms are arranged in ring structures. Examples include **benzene** and cyclic **alkanes** such as cyclohexane.

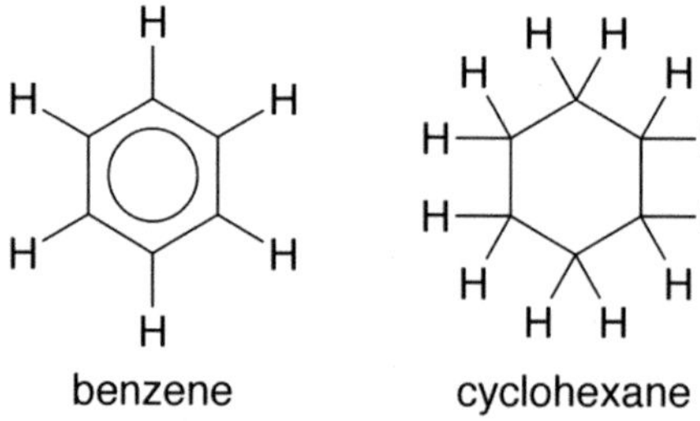

benzene　　　　cyclohexane

cyclic compounds

D

Daniell cell An electrolytic **cell** in which the following reaction takes place:

$$Zn(s) + Cu^{2+}(aq) \rightarrow Zn^{2+}(aq) + Cu(s)$$

dative bond *See* **coordinate bond**.

DC (direct current) The type of electrical **current** produced from a simple **cell** or **battery**. Only DC can be used in **electrolysis**.

DDT (dichlorodiphenyltrichloroethane) A widely used **pesticide** that has been successful in the control of diseases such as malaria. It is, however, harmful to animal life as it can accumulate in the body where it acts as a **poison**. For this reason many countries now restrict its use.

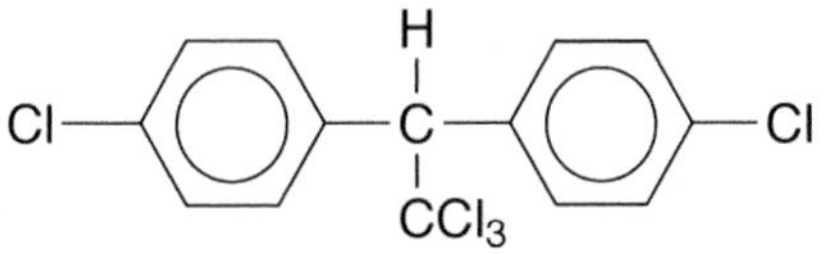

DDT

dead space *See* **burette**.

decomposition The breakdown of compounds into simpler compounds or into elements. Usually **heat** is needed (*thermal decomposition*), e.g.

$$2Cu(NO_3)_2(s) \rightarrow 2CuO(s) + 4NO_2(g) + O_2(g)$$

$$CaCO_3(s) \rightarrow CaO(s) + CO_2(g)$$

Electricity can also be used. *See also* **electrolysis**.

dehydrating agent Any substance that is used to remove water from other substances. Such substances always have an attraction for water. They may be substances that dissolve in water, such as concentrated sulphuric acid (H_2SO_4), sodium hydroxide (NaOH); or substances that react with water, e.g. calcium oxide (CaO) to form the hydroxide $Ca(OH)_2$.
   A third kind of dehydrating agent is the **anhydrous** salt which absorbs water, e.g. calcium chloride, sodium sulphate:

$$CaCl_2 + H_2O \rightarrow CaCl_2.2H_2O$$

$$Na_2SO_4 + H_2O \rightarrow Na_2SO_4.10H_2O$$

Dehydrating agents are used in **desiccators**.

dehydration   A chemical reaction that takes place by the removal of a molecule of water. The water molecule can be chemically combined as **water of crystallization**, or can come from chemical change within the molecule. An example is

$$CaSO_4.2H_2O(s) \rightarrow CaSO_4(s) + H_2O(g)$$

deliquescent substances   Substances that are able to take in water from the atmosphere. They can take in so much water that they are able to form a **solution** – unlike **hygroscopic** substances. Some common examples are iron(III) chloride ($FeCl_3$), copper(II) nitrate ($Cu(NO)_2$) and calcium chloride ($CaCl_2$).

Some deliquescent substances are used in **desiccators**.

delocalized electrons   Electrons that are free to move amongst the atoms. In normal **covalent bonds**, electrons are shared between two atoms, for example a carbon–carbon **single bond** consists of two electrons C:C, and a double bond contains four electrons C::C. These electrons are located in between the atoms, i.e. they are *localized*.

Graphite contains delocalized electrons within the hexagonal layers of atoms. This is why graphite can conduct electricity along the layers but not between them. Metals have delocalized electrons too and this explains why they are good **conductors**.

denaturing   The process by which **proteins** lose their three-dimensional structure. Proteins can be denatured by high temperature and by low and high **pH**.

density   A measure of how much space a certain amount of a particular substance takes up. The smaller the space into which the **mass** is concentrated, the greater is the density. The density of a substance is determined by the density of the individual atoms and also by how closely together the atoms are arranged in the crystal **lattice**. Generally speaking, the more particles that an atom contains, the denser it will be. Density is calculated by dividing a substance's mass by its **volume**. The units are usually either $g\,cm^{-3}$ or $kg\,m^{-3}$.

desiccator   A device used to store substances that need a dry atmosphere, e.g. **deliquescent substances** or **hygroscopic substances**. A desiccator is airtight and contains a **dehydrating agent** that makes the air dry.

desulphurization   The removal of compounds of sulphur from **petroleum**, **natural gas** or other **fossil fuels** before use. Desulphurization not only prevents the release of the acidic pollutant sulphur dioxide, but also allows the sulphur that is recovered to be sold for use. The sulphur is usually removed by using hydrogen and a **catalyst** to form a **hydrocarbon** and hydrogen sulphide (H_2S). The H_2S is then treated chemically to produce sulphur. A recovery rate of 95% is possible.

detergent, soapless detergent or **surfactant**
A cleaning agent which does two things when used in washing. It reduces the surface tension of the water and so allows the water to wet things more thoroughly. It also acts to bring together the water and (normally insoluble) fat, oil or grease into an **emulsion**. If the presence of a detergent is accompanied by rapid movement, e.g. in a washing machine, then particles of dirt and grit are removed and the article is cleaned.
   Detergents form an emulsion because one part of their molecule is **ionic** and is attracted to water while the other part (the **hydrocarbon** chain) is attracted to the oil molecules. Detergents thus hold the two together.

- 12%   detergent
- 40%   builders such as polyphosphates to remove calcium and magnesium ions from the water
- 12%   bleach such as sodium perborate
- 20%   anti-caking chemical to make sure that the powder runs smoothly
- 10%   water
- 5%   other substances such as dyes, enzymes, perfume
- 1%   suspension agent to keep the dirt suspended in the water

detergent *Contents of a typical packet of powder for an automatic washing-machine.*

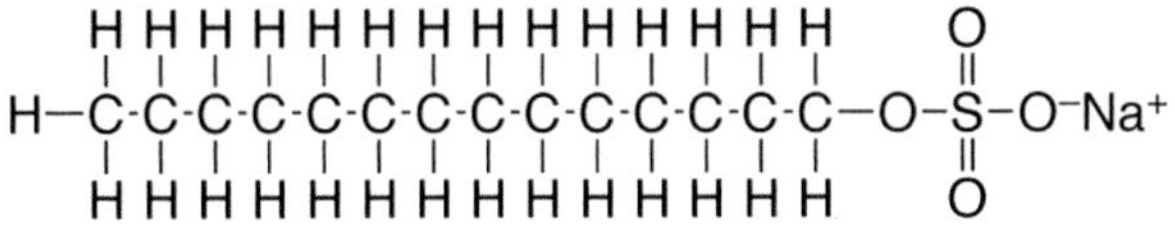

detergent *A synthetic (soapless) detergent molecule.*

See also **soap**; **hydrophilic**.

deuterium   An **isotope** of **hydrogen** which has a **neutron** and a **proton** in the **nucleus**. Its symbol is written

$$^2_1H \text{ or } ^2_1D$$

The isotope occurs naturally, making up 0.015% of hydrogen. Because its density is twice that of normal hydrogen it is easily separated.

Deuterium oxide is known as *heavy water* and its formula is written D_2O.

di- A prefix that means two. For example:
• Carbon dioxide CO_2 – two atoms of oxygen.
• diatomic molecule – two atoms in the molecule.

diamond A valuable mineral that is an **allotrope** of carbon. It occurs naturally, mainly in South Africa and the Russian Federation. It is the hardest naturally-occurring substance known and is used in drill tips and saw blades. Diamond is prized as a gemstone because of its rarity and the sparkle it produces. The carbon atoms are arranged in the crystal **lattice** in a **tetrahedral** formation and are covalently bonded to each other (*see* **covalent bond**).

diatomic molecule A **molecule** of an element that contains two atoms, N_2, O_2, H_2, Cl_2, etc.

diaphragm cell An electrochemical **cell** for the production of chlorine and sodium hydroxide from **brine**. In the cell, the **anode** and **cathode** are separated by a porous membrane, called a *diaphragm*, which prevents the chlorine formed at the anode from reacting with the sodium hydroxide formed at the cathode. This means of producing chlorine and sodium hydroxide is now coming back into greater use because it does not involve the use of mercury as a cathode in **electrolysis** cells.

dibasic acid An **acid** that contains two replaceable **hydrogen** atoms per molecule. An example is sulphuric acid H_2SO_4. With such acids, two kinds of salt can be formed – the normal salt, in which both hydrogen atoms are replaced, and the **acid salt** in which only one hydrogen has been replaced.

Acid	Salts	
	acid	*normal*
H_2SO_4	$NaHSO_4$	Na_2SO_4
sulphuric acid	sodium hydrogensulphate	sodium sulphate

dibasic acid

dichromate(VI) ion ($Cr_2O_7^{2-}$) An **ion** that contains chromium and oxygen atoms and has a valency of 6. It is usually used as the bright orange potassium or ammonium salts, $K_2Cr_2O_7$ and $(NH_4)_2Cr_2O_7$.

The dichromate ion is an **oxidizing agent** which is reduced to chromium(III) ions Cr^{3+}.

diesel fuel   A fuel/air mixture is injected into diesel engines where it is compressed. The temperature produced in the compression causes the fuel/air mixture to explode. The fuel is produced from **petroleum** and is made up mainly of **alkanes** in the boiling range 200–350°C.

diethyl ether   *See* **ethers**.

diffusion   The complete mixing of two or more different substances that comes about from the natural movements of the particles of the substance. Diffusion occurs rapidly in gases because the molecules in the gases are moving about randomly at high speed because of the **thermal energy** they possess.

Diffusion occurs at a faster rate when the temperature is raised. The less dense a gas is, the greater is its rate of diffusion (*see* **density**).

Diffusion occurs at a slower rate in liquids and solutions and yet more slowly in solids.

dilute solution   A **solution** that contains a relatively low **concentration** of **solute**. If you want to make a solution more dilute it is usual to add more solvent to it. Dilute acids usually have concentrations of 2 mol dm^{-3} or less.

dimer   A **molecule** which is made up of two identical molecules (**monomers**) which are bonded together.

$$\underset{O}{\overset{O}{\diagdown}}N-N\underset{O}{\overset{O}{\diagup}}$$

dimer N_2O_4 *as a dimer of* NO_2.

disaccharide   A **sugar** molecule that is made up of two **monosaccharide** sugar molecules that have undergone a **condensation** reaction with the elimination of a molecule of water.

Monosaccharides	Disaccharide
glucose + glucose	maltose
glucose + fructose	sucrose

disaccharide *Monosaccharide sugars that react to form disaccharides.*

displacement reaction   A reaction in which a less reactive element is displaced by a more reactive one. These are **redox** reactions. For example:

$$CuSO_4(aq) + Zn(s) \rightarrow ZnSO_4(aq) + Cu(s)$$

$$2NaBr(aq) + Cl_2(g) \rightarrow 2NaCl(aq) + Br_2(aq)$$

dissociation The breaking up of a compound into smaller simpler molecules or ions. The dissociation is sometimes **reversible**.

dissolve To become mixed with and absorbed into a liquid. When a **solute** dissolves in a **solvent** to form a **solution** it is dispersed throughout the whole volume of the solvent. The structure of the solute is broken up.

$$\underset{\substack{\text{ammonium} \\ \text{chloride}}}{NH_4Cl(s)} \quad \overset{\text{heat}}{\underset{\text{cool}}{\leftrightarrows}} \quad \underset{\text{ammonia}}{NH_3(g)} \quad + \quad \underset{\substack{\text{hydrogen} \\ \text{chloride}}}{HCl(g)}$$

distillation The process by which a **solvent** can be recovered from a **solution** or a **mixture**. In a simple distillation, the solution is heated and the solvent is turned into a vapour. The vapour is led away from the hot flask into a **Liebig condenser** where it cools and condenses to a liquid, and is collected. In this way pure solvent can be recovered. *See also* **condensation**, **fractional distillation**.

distilled water Water that has been treated by **distillation** to increase its purity. Tap water and rain water are not pure. They contain dissolved **salts** and gases. In distillation, most of the salts are left behind but the water still may contain dissolved gases. The presence of carbon dioxide reduces the **pH** of the water considerably.

dm³ (cubic decimetre) One thousand cubic centimetres. *See* **cm³**.

dolomite A common mineral containing calcium carbonate ($CaCO_3$) and magnesium carbonate ($MgCO_3$). Dolomite is an important source of magnesium and is used in furnace linings.

double bond A **covalent bond** that contains two *shared pairs* of electrons. Carbon–carbon double bonds are found in **alkene** molecules. The presence of the extra electrons makes the **alkenes** very reactive. Such compounds which contain double bonds (and **triple bonds**) are said to be **unsaturated** compounds. Examples are shown below:

Oxygen $O=O$

Ethene $\underset{H}{\overset{H}{\diagdown}} C = C \underset{H}{\overset{H}{\diagup}}$

Carbon
dioxide $O=C=O$

double decomposition or precipitation reaction A process that occurs when ionic substances react in solution. The product is an insoluble solid. For example:

$$CaCl_2(aq) + Na_2CO_3(aq) \rightarrow CaCO_3(s) + 2NaCl(aq)$$

or

$$Ca^{2+}(aq) + CO_3^{2-}(aq) \rightarrow CaCO_3(s)$$

This is a method of preparing insoluable **salts**.

Dow process A process for the extraction of magnesium from brine using calcium hydroxide solution to precipitate magnesium hydroxide, which is then dissolved in hydrochloric acid, and the solution **electrolysed**.

dry cell An electrolytic **cell** in which the **electrolyte** is in the form of a paste and so cannot spill from the battery. The only common *wet* battery used today is the lead–acid **accumulator**.

dry cleaning A process for cleaning fabrics that cannot safely be washed using water. The cleaning agents used are **hydrocarbon** compounds containing chlorine.

dry ice Solid carbon dioxide. This material sublimes (*see* **sublimation**) so no melting takes place and it is described as dry. Dry ice is used for keeping things cold as its sublimation point is –40°C, and is also used in the theatre for producing smoke and cloud effects on stage.

drying agent A substance used to extract the water from another material. Drying agents can also be used to keep substances dry and will often be packaged inside devices such as cameras. Examples are calcium oxide (CaO) and **silica gel**. *See also* **dehydrating agent**.

ductile Easily shaped. A ductile substance can easily be drawn into wires; copper, for example, is a very ductile metal. The term is usually used only to describe metals. Ductile metals have large **crystals** within them. *See also* **annealing**, **malleable**.

dyes Chemicals that can be mixed or reacted with materials to make a coloured product. The colour is produced because the dye absorbs some of the light that falls upon it and radiates the rest. *See also* **pigment**.

dynamic equilibrium *See* **equilibrium**.

dynamite   An explosive invented by Alfred Nobel (1833–96). He discovered that if the unstable explosive nitroglycerine was absorbed into a clay called *Kieselguhr* the result was a stick of explosive that was safe until it was detonated. *See also* **explosion**.

E

Earth's atmosphere   The mixture of gases surrounding the Earth that protects us from harmful radiation from the Sun and deep space, and provides part of the means of sustaining life on the planet. On Earth, this mixture of gases is called air.

The **atmosphere** is more than 1000 km deep, although over 90% of the air is found in the first 17 km. It is made up of several layers: *thermosphere*, *mesosphere*, *stratosphere* and *troposphere*. *See also* **global warming**, **greenhouse effect**.

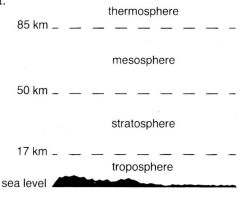

Earth's atmosphere *The layers making up the atmosphere.*

Earth's structure   The Earth is made up of the *crust*, the *mantle* and the *core*. The crust is a relatively thin layer that varies in thickness from about 7 km under the sea to about 40 km under the continents; the top metre or so may be soil. The mantle is about 2870 km thick and is made up of a molten layer held between two more rigid layers. The core is about 3500 km thick and is made up mainly of iron and nickel.

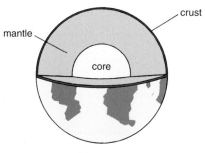

Earth's structure

efflorescence The process by which **crystals** lose part of their **water of crystallization** when they stand in the air. They go powdery on their surfaces. A well known example of this process is sodium carbonate (washing soda):

$$Na_2CO_3.10H_2O(s) \rightarrow Na_2CO_3(s) + 10H_2O(g)$$

elastomer A substance (usually a **polymer**) with elastic properties, i.e. the ability to stretch and return to its original shape. Examples are synthetic and natural **rubber.**

electric arc process *See* **steel manufacture.**

electric current or electricity *See* **current.**

electrochemical reactions *See* **cell**, **electrolysis**, **electroplating**.

electrochemical series A 'league table' of elements arranged in order of their readiness to release electrons and form cations. Elements differ in their reactivity.

K	potassium	
Na	sodium	
Ca	calcium	
Mg	magnesium	
Al	aluminium	
Zn	zinc	reactivity *increases*
Fe	iron	
Pb	lead	
H	hydrogen	
Cu	copper	
Ag	silver	

electrochemical series

When a metal is dipped into a solution of one of its salts, e.g. zinc in zinc sulphate solution, the following reaction occurs:

$$zinc \rightarrow zinc\ ions + 2\ electrons$$

$$Zn(s) \rightarrow Zn^{2+}(aq) + 2e^-$$

The **ions** go into solution and the electrons cling to the metal. If the metal and solution are made part of a circuit the electrons will flow round the circuit. The more reactive a metal is the greater the force with which the electrons move.

   The series is useful because it allows us to compare reactivities and hence predict **reactions**, e.g.:

(a) Zinc will remove the oxygen from copper(II) oxide, but
(b) Copper will not remove oxygen from zinc oxide;
(c) Hydrogen will reduce copper(II) oxide but not zinc oxide;
(d) Copper will not react with acids to release hydrogen.

electrode A conductor that dips into an **electrolyte** and allows the current (electrons) to flow to and from the electrolyte. Typical electrodes are made of copper, carbon and platinum. In **electrolysis** chemical reactions occur at the electrodes. *See also* **anode**, **cathode**.

electrolysis A process in which a direct electric current (**DC**) is passed through a liquid that contains **ions** (an **electrolyte**) to produce chemical changes at the **electrodes**. Electrolysis is used to extract metals and non-metals from compounds. Aluminium, sodium, magnesium, chlorine and copper are examples of elements produced in this way. For example:

<table>
<tr><th>At the anode</th><th>At the cathode</th></tr>
<tr><td>$Cl^- \rightarrow Cl + e^-$</td><td>$H^+ + e^- \rightarrow H$</td></tr>
<tr><td>$2Cl \rightarrow Cl_2$</td><td>$2H \rightarrow H_2$</td></tr>
<tr><td>chlorine is given off</td><td>hydrogen is given off</td></tr>
</table>

electrolysis *The electrolysis of hydrochloric acid.*

electrolyte A liquid that conducts electricity as a result of the presence of negative or positive ions. An electrolyte may be either a molten ionic compound or a **solution** containing **ions**.
   Chemical changes take place when a direct electric current is passed through an electrolyte (**electrolysis**). With molten compounds such as sodium chloride the changes are simple:

$$2Na^+Cl^-(l) \rightarrow 2Na(l) + Cl_2(g)$$

The elements are produced.
   With ionic solutions, the processes are more complex. The products depend on the **concentration** of the solution, the voltage which is applied, the type of **electrode** used, and the nature of the ions present in the solution.

electrolytic refining A method of purifying metals. An impure piece of

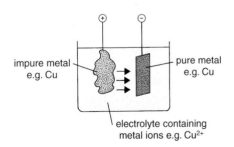

impure metal
e.g. Cu

pure metal
e.g. Cu

electrolyte containing
metal ions e.g. Cu^{2+}

electrolytic refining

metal is made the **anode** of an electrolytic **cell**, with a thin sheet of pure metal as the **cathode**. The **electrolyte** contains the metal ions. During **electrolysis**, metal is transferred from anode to cathode through the electrolyte.

electron   One of the basic particles of matter. Its mass is around $\frac{1}{1840}$ of that of the **proton**; it carries a negative **charge**. There are electrons in all normal **atoms**. They are given the symbol e^-.

Atoms form **ions** through the loss or gain of electrons. An electric **current** is electrons moving through a **conductor**. The way in which electrons are located in an atom is known as the **electronic configuration**.

electronegativity   The power that an atom has to attract **electrons** to itself in a chemical **bond**. As the diagram shows, **fluorine** has the most electronegative atoms.

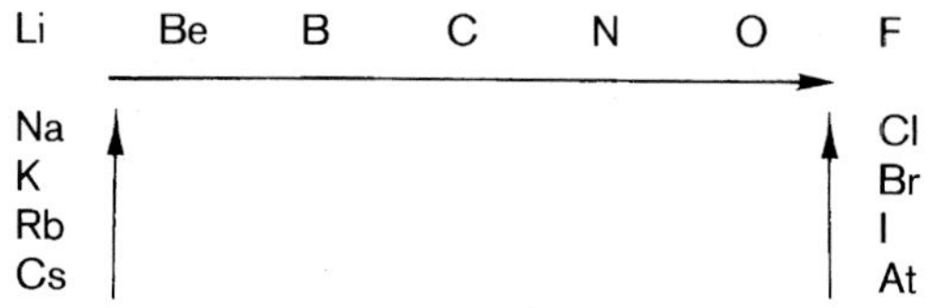

Li	Be	B	C	N	O	F
Na						Cl
K						Br
Rb						I
Cs						At

electronegativity *Electronegativity increases across a period and up a group.*

electronic configuration   The arrangement of **electrons** as they move around the **nucleus** of an **atom**. The movement takes place in clearly-defined regions (*orbitals*) in the atoms. Electrons may be visualized as being arranged in orbits, or what are called shells, each of which can contain a maximum number of electrons:

Shell number:	①	②	③	④	⑤
Maximum number of electrons:	2	8	18	32	50

Across the **periodic table**, the number of electrons increases by one per atom as the **atomic number** increases. The shells are filled up in turn – ① first, then ②, and so on – so that the electrons go into the region of the atom that has the lowest energy. For some elements described in this book, the electronic configuration is given, e.g. sodium: 2·8·1. In other words, the ① and ② shells are full and the ③ shell is beginning to fill.

The periodic table is arranged according to electronic configuration.

Thus those elements with an electronic configuration ending in 1 make up the group I elements:

| Lithium | 2·1 |
| Sodium | 2·8·1 |

Those that end in 7 (the group number) make up the **halogens:**

Fluorine	2·7
Chlorine	2·8·7
Bromine	2·8·18·7

electrophoresis   The movement of colloidal particles (*see* **colloidal dispersion**) in an electric field. Different types of particle move in different directions and at different speeds; hence electrophoresis can be used to identify substances.

electroplating   A method of coating one metal with a thin layer of another metal. The metallic object to be coated is made the **cathode** and undergoes **electrolysis** in a bath where the **electrolyte** contains ions of the coating metal. The current, electrolyte concentration and temperature must be carefully controlled, as they affect the deposited metal.

Electroplating is usually carried out either for protection, e.g. chromium plating on bicycle handlebars and motor cars, and tin plating on cans; or for decoration, e.g. silver plating on cutlery and ornaments, and gold plating on jewellery, etc.

electrostatic precipitation   A method of removing dust from flue gases and other waste products. Gases are passed through a chamber that has a positive electrical charge (about 10 kilovolts). The negatively charged dust particles are attracted to the sides of the chamber, leaving the gas clean.

electrovalent bond   *See* **ionic bond**.

element   A pure substance that cannot be broken down into anything simpler by chemical means. There are 118 elements known to us. Most of these occur naturally on the Earth but several have been made in laboratories by **nuclear reactions**. All elements have a unique number of **protons** in their atoms. There are three broad classes of element:
(a) **metals**, e.g. iron;
(b **non-metals**, e.g. oxygen;
(c) **metalloids**, e.g. arsenic.

elementary particle   The subatomic particles found in the **atom**. They are the **proton**, **neutron** and **electron**.

emery   Aluminium oxide (Al_2O_3). It is used as an **abrasive**.

empirical formula   The formula of a compound that shows the atoms that are present in the molecule in their simplest ratio. It is not possible to tell from the empirical formula whether a compound is a **molecule** or a **giant structure**. *See also* **molecular formula, percentage composition**.

Compound	Molecular formula	Empirical formula
Ethene	C_2H_4	CH_2
Butane	C_4H_{10}	C_2H_5
Propane	C_3H_8	C_3H_8
Ethanoic acid	$C_2H_4O_2$	CH_2O

empirical formula *Comparisons of the empirical and molecular formula of compounds. Note that the formulae for propane are the same.*

emulsifier   A substance that prevents an **emulsion** separating into layers. There are many natural emulsifiers; these and artificial **additives** are widely used in foods, cosmetics and medicines.

emulsion   A **colloidal dispersion** of a liquid within a liquid that does not separate into layers. Milk is an emulsion with fat particles spread throughout the liquid. Salad cream and mayonnaise are emulsions of oil in vinegar. A natural **emulsifier** in egg yolk helps to keep the emulsion stable.

enamel   A thin glass coating that is applied to materials such as metals to provide a protective and decorative coating. *See also* **ceramic**.

endothermic reaction   A reaction in which **heat energy** is taken in from the surroundings. There is either a fall in temperature in the reactants when the reaction occurs, e.g. when sodium nitrate dissolves in water:

$$NaNO_3(s) \rightarrow NaNO_3(aq)$$

or heat energy has to be continually supplied to make the reaction occur, e.g.:

$$CaCO_3(s) \rightarrow CaO(s) + CO_2(g)$$

In endothermic reactions, there is more energy in the bonds of the products than there was in the bonds of the reactants. *See also* **enthalpy change diagram**.

end point   The point at which a reaction is complete. The end point of a **titration** is when one of the reactants has been completely used up.

The end point can be identified by an **indicator** or an instrument such as a **pH** meter or a conductivity meter.

energy   The 'ability to do useful work'. Energy is 'locked' inside the nuclei of atoms. Sometimes this can be released (*see* **nuclear reactions**). Energy is also found in the bonds between atoms. When chemicals react, bonds break and new ones form. In this breaking and forming, energy can be absorbed, or it can be released as heat energy (*see* **energy change**). In batteries, however, it is released as electrical energy. Energy is usually measured in **joules** or **kilojoules**. *See also* **enthalpy**.

energy change   The transfer of **heat energy** during a chemical reaction. Heat energy can be either released from the reactants (*see* **exothermic reaction**); or it can be taken in by the reactants (*see* **endothermic reaction**).
   Energy changes are measured in **joules** or kilojoules and relate to a particular amount of substance, usually a **mole**. Thus the energy change for the **Haber process** is –92 kilojoules per mole (–92 kJ mol^{-1}). The negative sign indicates that heat energy is released into the surroundings.

enthalpy   A measure of the **heat energy** possessed by a chemical substance. It is given the symbol H. It cannot be directly measured but, when chemicals react, the difference in the heat energy between the reactants and products can be measured by using a thermometer. This difference is the enthalpy change and is given the symbol ΔH. ΔH is negative for an **exothermic reaction** and positive for an **endothermic reaction**.

enthalpy change diagram   A diagram that shows the change in **enthalpy** for two chemical reactions (A→B; A→C). A→B is an **endothermic reaction**, as product (B) has a greater enthalpy at the end that at the start; A→C is an **exothermic reaction**, as product (C) has a lower enthalpy at the end.

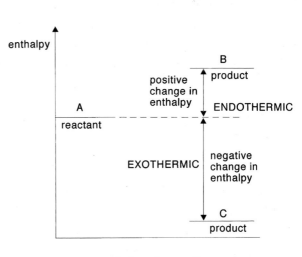

enthalpy change diagram

enzymes   A natural **catalyst**; a biological molecule that changes the rate of a reaction without undergoing a chemical change itself. Enzymes are found in living tissue and allow complicated biochemical reactions to occur at low temperatures and pressures inside the body. They are generally **protein** molecules. It is thought that an enzyme is a particular shape that allows two reacting molecules to come close together on the enzyme. This makes the reaction easy to carry out.

epoxy resins   A group of strong, inert synthetic **polymers** that are electrical insulators. They have wide uses as **adhesives**, coatings and **composite** materials. The resins are produced by mixing a poly(ether) with a chemical reagent.

equation   (in chemistry) A way of describing a reaction. The equation can be a *word equation*:

$$hydrogen + oxygen \rightarrow water$$

or it can be a *formula* (or symbol) *equation*:

$$H + Cl \rightarrow HCl$$

An equation is said to be *balanced* if there are the same number of each kind of atom on each side of the equation. For example,

$$2H_2 + O_2 \rightarrow 2H_2O$$

is a balanced equation. Two molecules of hydrogen react with one molecule of oxygen to produce two molecules of water. This equation also tells us that 2 moles, or 4 g, of hydrogen react with 1 mole or 32 g of oxygen to produce 2 moles, or 36 g, of water. Since the relative atomic mass of hydrogen is 1 and that of oxygen is 16:

$$2H_2 + O_2 \rightarrow 2H_2O$$
$$2(2) + 32 = 2(2 + 16) = 36$$

Sometimes **ionic equations** are used to represent a reaction:

$$Cu^{2+} + Zn \rightarrow Cu + Zn^{2+}$$

State symbols are usually added after the formula to denote which **state of matter** the substance is in. For example, gaseous hydrogen and oxygen react to produce liquid water:

$$2H_2(g) + O_2(g) \rightarrow 2H_2O(l)$$

equilibrium A state of balance or rest. In a **reversible reaction** the reaction proceeds in both directions, for example

$$3Fe(s) + 4H_2O(g) \rightleftharpoons Fe_3O_4(s) + 4H_2(g)$$

The reactants (iron and steam) produce the oxide and hydrogen. As soon as the products are made, they begin to react to form steam and iron again. There eventually comes a time when the **rate** of the forward $\rightarrow$ reaction equals the rate of the back $\leftarrow$ reaction. At this point the proportion of substances present is constant. There seems to be no reaction taking place. There is an equilibrium between the forward and back reactions. It is important to realize that the forward and back reactions are still occurring but, because they are taking place at the same rate, no change can be seen. Such an equilibrium is termed *dynamic*, that is, it produces motion.

   Many important reactions involve equilibria. In the **Haber process**, for example, with the usual operating conditions, less than 20% of the hydrogen and nitrogen are converted into ammonia. This is the equilibrium **yield** of ammonia in the reaction.

essential amino acid *See* **amino acid**.

essential fatty acid A **fatty acid** that is needed in the human diet because the body does not produce any or enough of it. There are four main essential fatty acids, all of which are found in vegetable oils.

ester A compound formed between an alcohol and a **carboxylic acid**. A strong acid **catalyst** is usually needed for the reaction. The reaction between acid and alcohol is known as *esterification*.

   The ester group has the following combination of atoms:

$$R^1 - C \underset{O-R^2}{\overset{O}{\|}}$$ where R^1 and R^2 are hydrocarbon chains.

   Esters often have sweet, fruit smells and are used in perfumes, flavourings and essences.

ethane An **alkane**. It is a colourless, flammable gas that is found in all **natural gas**. **Ethene** is formed when ethane is subjected to **cracking**.

C_2H_6

ethane

ethanoic acid (formerly **acetic acid**) A colourless liquid (melting point 16 °C, boiling point 118 °C).

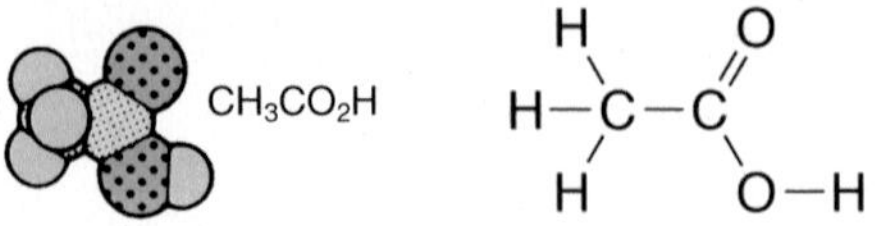

CH₃CO₂H

H—C—C with H, H, O, O—H structure

ethanoic acid

It is a **carboxylic acid** that is made industrially by the oxidation of **naphtha**. It is used to produce **esters** and other chemicals which are used to make **polymers** from which fabrics are manufactured.

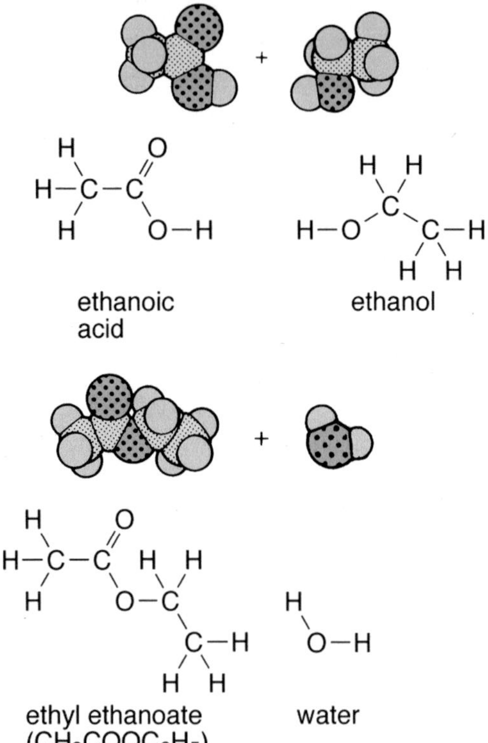

ethanoic
acid

ethanol

ethyl ethanoate
($CH_3COOC_2H_5$)

water

ethanoic acid *Reaction with ethanol.*

Ethanoic acid is present in **vinegar** (about a 4% aqueous solution). Although it is a **weak acid**, in a concentrated form it can cause severe skin burns.

ethanol A colourless, flammable liquid **alcohol** whose boiling point is 78 °C. Ethanol is the alcohol contained in alcoholic drinks. These are made by **fermentation**, but in the chemical industry pure ethanol is produced by the hydration of **ethene**:

$$C_2H_4(g) + H_2O(l) \rightarrow C_2H_5OH(l)$$

Ethanol is used in industry as a solvent, to produce **esters** and **ethers**, and to make cosmetics. It is also used in **methylated spirits**. It can be oxidized to ethanal and **ethanoic acid**, but this is now not carried out commercially. Ethanol can also be used as a fuel.

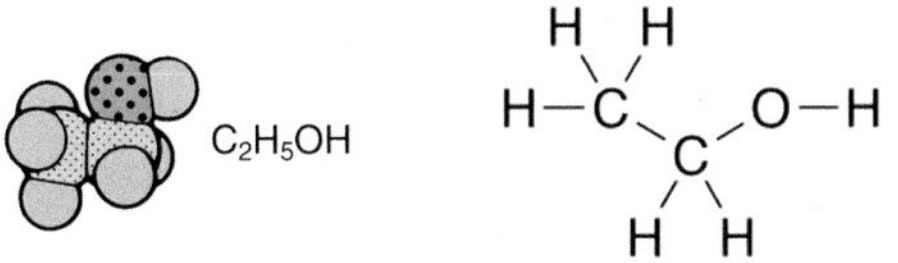

ethanol

ethene (formerly ethylene)
A gaseous **alkene** that is produced by the **cracking** of alkanes (e.g. **ethane** and **naphtha**). Because it is an **unsaturated** compound, ethene is reactive. It is a very important chemical, being used to make plastics such as **poly(ethene)** and **poly(chloroethene)**. Ethene can add molecules across its **double bond**, for example to produce **ethanol**. These are **addition reactions**.

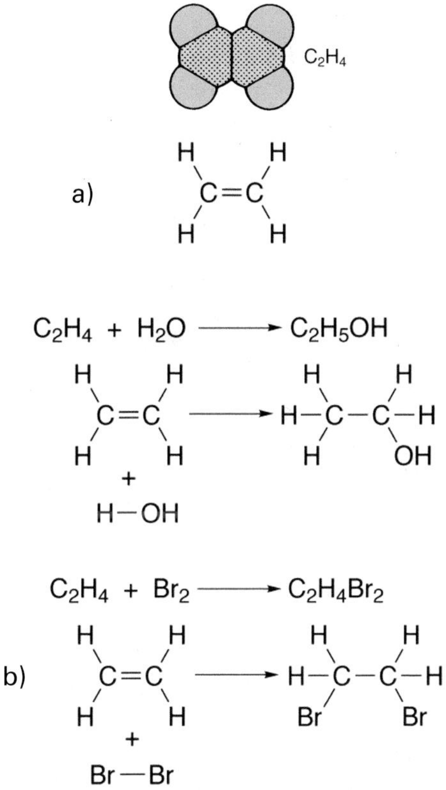

ethers   A group of organic compounds that have an oxygen atom bonded to two carbon atoms. One of the most important ethers is diethyl ether (ethoxyethane).
   Diethyl ether is usually known simply as *ether* and was once widely used as an **anaesthetic**. Nowadays it is used as a solvent for substances that do not dissolve in water. It is very

ethene (a) The compound. (b) Adding molecules across its double bond to produce ethanol and ethylene dibromide.

flammable and air/ether mixtures are dangerously explosive. Diethyl ether is produced by the action of concentrated sulphuric acid on ethanol.

$$C_4H_{10}O$$

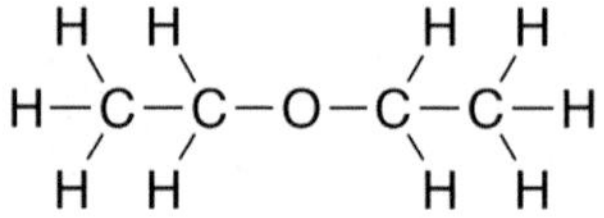

H—C—C—O—C—C—H (with H atoms above and below each carbon)

ethers *Diethyl ether.*

ethylene   *See* **ethene.**

ethyl ethanoate   The **ester** produced by reacting **ethanol** and **ethanoic acid**. It is used as a solvent in glues and paints.

ethyne or acetylene   A flammable gaseous **alkyne** that forms explosive mixtures with air. It can be produced by the **cracking** of petroleum products, although it used to be made by the action of water on calcium carbide:

$$CaO(s) + 3C(s) \rightarrow CaC_2(s) + CO(g)$$

$$CaC_2(s) + 2H_2O(l) \rightarrow C_2H_2(g) + Ca(OH)_2(s)$$

Ethyne gas was once widely used for illumination, being burnt in special lamps that were used in mines and on bicycles. It was also used feedstock for **chloroethene** and **poly(chloroethane)**, but has has largely been replaced by other reagents because of its cost. Ethyne produces very high temperatures when burnt in an oxygen-enriched gas supply, and is used in this manner in the oxyacetylene torch for welding and cutting metal. Temperatures of 3000 °C can be obtained.

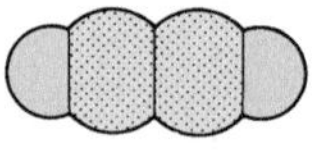

$$C_2H_2$$

$$H-C{\equiv}C-H$$

ethyne

eutrophication   The process by which the water in lakes and rivers become over-rich in nutrients. Eutrophication is caused by fertilizers and other chemicals such as **phosphates** from sewage works. These get into the water and cause the rapid growth of unwanted plants that choke the water and remove much of the oxygen; it also prevents sunlight penetrating down into the water. A lake or river is described as *eutrophic* if there is not enough oxygen in the water to support animal and plant life. *See also* **acid rain, pollution**.

evaporation The process leading to a change of **state** from liquid to vapour which can occur at any temperature up to the **boiling point**. It takes place because molecules escape from the body of the liquid into the atmosphere. Only a small proportion of the molecules have sufficient energy to escape at any time but over a period they will all escape. Generally speaking, the lower the boiling point, the faster the rate of evaporation.

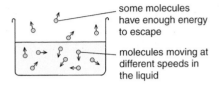

some molecules have enough energy to escape

molecules moving at different speeds in the liquid

evaporation *Molecules escaping from the body of a liquid into the atmosphere.*

exothermic reaction A reaction in which **heat energy** is released to the surroundings from the reactants. The chemical bonds of the products contain less energy than the bonds of the reactants. The products are more stable than the reactants.

Many common reactions are exothermic. All **combustion** and **neutralization** reactions are exothermic. For example:
(a) the **Haber process** $3H_2(g) + N_2(g) \rightarrow 2NH_3(g)$
(b) the **contact process** $2SO_2(g) + O_2(g) \rightarrow 2SO_3(g)$
are both exothermic reactions at **room temperature**. **Enthalpy** changes (ΔH) are less than zero (negative) in exothermic reactions. *See also* **endothermic reaction, energy change, enthalpy change diagram**.

explosion A rapid expansion of gas that can cause damage. In the reaction between an explosive substance and oxygen, chemical energy is released, resulting in sudden very high temperatures and the production of a large quantity of gas. The gas expands very rapidly in the heat, producing a *shock wave* (pressure wave) that travels outwards from the explosion, and it is the energy in the shock wave that causes the damage in explosions. Common explosives are gunpowder, **dynamite**, nitroglycerine, TNT, and plastic explosives such as Semtex. In nuclear explosions the vast amount of energy locked within the nucleus of atoms is released.

F

faraday constant (F) The amount of electric charge possessed by a **mole** of electrons (about 96,500 **coulombs**). This amount of electricity is needed to liberate one mole of atoms of a *univalent* element, i.e. one (e.g. silver, Ag) that releases one electron to form its cation during **electrolysis**:

$$Ag^+ \quad + \quad e^- \quad \rightarrow \quad Ag$$

One mole of silver ions.   One mole of electrons.   One mole of silver atoms.

fats or lipids **Esters** of **fatty acids**. They are common constituents of food and a good source of energy, but there are worries that the UK diet contains too much fat. Oils are fats which are liquid at 20 °C.

fatty acid An organic **acid** with the general formula $C_nH_{2n}O_2$. These compounds are found in both animals and plants, combined with glycerol as **esters**. They form the oils and **fats** that are such an important part of our diet.

Fatty acids can be **saturated compounds**, where all the carbon–carbon bonds are single bonds, or **unsaturated compounds**, where some of the carbon–carbon bonds are double ones.

Fehling's test A test used to detect certain organic **reducing agents**. **Sugars** such as **glucose** can be distinguished from **starch** by using the test. Fehling's solution is a mixture of copper(II) sulphate, sodium hydroxide and sodium potassium tartrate. When it is heated with an appropriate reducing agent, such as **glucose**, the copper(II) salt is reduced to a red **precipitate** of copper(I) oxide (Cu_2O).

fermentation A process whereby chemical changes are made to organic chemicals by the use of living organisms such as yeasts and bacteria. The changes are brought about by **enzymes** acting as **catalysts**.

One important example is the changing of **sugars** to **ethanol** by the action of yeast:

$$\overset{\text{zymase}}{C_6H_{12}O_6 \quad \rightarrow \quad 2C_2H_5OH \quad + \quad 2CO_2}$$

glucose   ethanol   carbon dioxide

This reaction is used to make alcoholic drinks:

ferroxyl
indicator A test for Fe^{2+} ions. *See* **iron**.

fertilizer A substance that is added to the soil to replace nutrients that have been lost, or removed by plants. Fertilizers may be natural, for example manure and compost, or artificial, for example ammonium nitrate, superphosphate. Compounds containing nitrogen, phosphorus and potassium (*NPK*) are the main constituents of artificial fertilizers. The addition of fertilizers to the land is an important part of the **nitrogen cycle**.

Ammonium (NH_4^+) and potassium (K^+) ions are widely used in fertilizers because they are strongly attracted to negatively charged particles (e.g. clay) in the soil.

Overuse of fertilizers can have detrimental effects on the environment. **Nitrates** in the soil can be washed out (*leached*) and can enter waterways, lakes and the like. If this happens, they can enter the public water supply where high concentrations can be harmful, especially to the young. High **phosphate** concentrations in water lead to **eutrophication**, but most of the phosphate in rivers comes from sewage and effluent from washing powders rather than from fertilizers. *See also* **eutrophication, pollution, superphosphate**.

fibre **1.** A material that can be made from a yarn and used to produce fabrics by knitting or weaving. Fibres can be produced from natural materials such as wool, silk, cotton and linen, or from **synthetic** materials such as **nylon, poly(propene)** and **rayon**.
2. Bulky material from plants that swells inside the body and aids the digestion process. Fibre is an important part of our diet. Fruit and grain are good sources.

fibre glass Material made up of fine threads of **glass**. These can be formed into matting or into yarn. Fibre glass is used as an insulating material.

fibre-reinforced plastics **Composite materials** in which plastics are combined with fibres to create lightweight, strong materials with special properties. Fibres may be **glass**, carbon, silicon carbide, **poly(ethene)**, Kevlar.

filter 1. A device that allows some substances to pass through it but not others.
2. To pass something through a filter. An insoluble solid and a liquid can be separated by pouring the mixture into a filter paper in a filter funnel. The liquid (or solution) passes through the paper. This liquid is known as the *filtrate*. The solid (known as the *residue*) is left in the filter paper. The whole process is known as *filtration*.

fire triangle A **symbol** summarizing the three things that have to be present for a fire to burn: fuel, heat and oxygen.

fission The process of breaking or splitting into parts. If a molecule is split into two or more parts it is said to undergo fission.
    In **nuclear reactions** the fission of a large atom, for example uranium into two smaller ones, such as barium and krypton, releases enormous amounts of energy.

fire triangle

fixing nitrogen Any process that converts atmospheric nitrogen into compounds which are useful as **fertilizers**. Fixing occurs in the **Haber process** and is also carried out by nitrifying bacteria found on the roots of certain plants, such as peas and beans (legumes).

flame Light and heat energy given off when substances react and hot gases are produced. Flames vary in their colour and their temperature depending on the substance being burnt.
    In a gas **burner**, the flame produced when the air hole is closed is yellow because it contains tiny glowing particles of unburnt carbon. The flame produced when the air hole is open (*shown in the diagram*) has different parts to it. The temperature varies in the different parts of the flame.

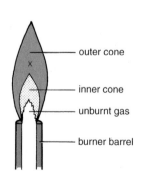

outer cone

inner cone

unburnt gas

burner barrel

flame *Appearance of a gas burner flame when the air hole is open. X marks the hottest place.*

flame test or **flame colour test** A test used to identify the metal ions present in compounds. A clean nichrome or platinum wire is dipped into concentrated hydrochloric acid and then into a sample of the compound under test. The wire, which now has a small amount of the compound attached to it, is then put into the hottest part of the **flame**. Metals are characterized by specific colours:

Metal	flame	Metal	flame
Copper	green/blue	Potassium	lilac
Calcium	brick red	Barium	apple green
Sodium	yellow/orange	Lithium	bright red

flax Natural **fibres** of **cellulose** that are used to make linen.

fluoride A negatively charged **ion** formed from **fluorine** by the gain of one electron. Fluoride compounds are added to toothpaste to reduce tooth decay and uranium hexafluoride (UF_6) is used to separate uranium **isotopes**. *See also* **poly(tetrafluoroethene)**, **fluothane**.

fluorine (F) A gaseous, non-metallic element in group VII of the **periodic table**. It is a **halogen** and the most reactive element known.

Fluorine is a vigorous **oxidizing agent** which will even oxidize **chloride** ions and water:

$$2Cl^-(aq) + F_2(g) \rightarrow 2F^-(aq) + Cl_2(aq)$$

$$2H_2O(l) + F_2(g) \rightarrow 4HF(g) + O_2(g)$$

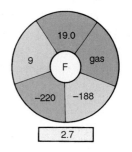

fluorine

Compounds of carbon and fluorine (**CFCs**) are important refrigerants and **aerosol** propellants.

Fluorine is extracted by the **electrolysis** of a molten fluoride, such as potassium hydrogen fluoride, KHF_2.

fluothane The most widely used **anaesthetic**, introduced in 1956. It is a colourless liquid with a pleasant smell; it is non-toxic and non-flammable.

foam A **colloidal dispersion** of a gas within a liquid, in which small bubbles of gas are separated by a thin film of the liquid. Foams will not form if the substances are pure, as stabilizers are needed. In ice creams and other foods where foams are used, **additives** are used to hold the foam together. Foams are used in fire-fighting because they are good at smothering the burning material and preventing flammable vapours from escaping. Solid foams are also found, e.g. pumice, polyurethane foam, meringue.

formaldehyde *See* **methanal**.

formalin An **aqueous solution** of **methanal**. It is used to preserve biological specimens.

formula *See* **empirical formula**, **molecular formula**.

forward reaction *See* **equilibrium**.

fossil The remains or evidence of a plant or animal that has been preserved in rock by natural processes.

fossil fuel A **hydrocarbon** that has been formed below ground from organic substances, and that can be extracted and burnt as **fuel**.

fractional distillation or **fractionation** A process of separating the components of a **mixture**. A **solvent** can be separated from a **solution** by the process of simple **distillation**. For a mixture of liquids with different boiling points it is usual to distil it in a *fractionating column* and collect the products (*distillates*) which boil in definite temperature ranges. These distillates are termed *fractions*.

 The exact temperature range of these fractions can be varied depending on the product which is in greatest demand. The production of whisky and brandy also involves fractional distillation. *See also* **fuel**.

fractionating column *See* **fractional distillation**.

Frasch process A method of obtaining **sulphur** from deep underground. A double well is sunk and superheated water is passed down. This liquifies the sulphur which is then blown to the surface by compressed air.

freeze-drying The removal of water from a frozen substance by evaporation at low pressure. This is used in the food industry to produce instant coffee and elsewhere in industry to process sensitive material which might be damaged by other methods of processing.

freezing or **solidification** The process by which a liquid becomes a solid. *See also* **kinetic theory**, **melting**.

freezing point The temperature at which a liquid becomes a solid. *See also* **melting point**.

fructose A **monosaccharide** sugar with the formula $C_6H_{12}O_6$. It is found in fruit and honey.

fuel A substance that releases **heat energy** when it is treated in a certain way. In most fuels, the energy is released by **combustion**. Nuclear fuels, such as uranium and plutonium, produce heat because of changes that occur inside the atom. The **nuclear reactions** generate very large amounts of energy.

Fossil fuels	Organic fuels	Energy release
Natural gas	Wood	All burn in oxygen
Petroleum	Waste materials	to release carbon
Coal	Biomass	dioxide and water
Peat		

fuel *Releasing heat energy through combustion.*

fuel cell A device that converts chemical energy directly into electrical energy. The way it works is the reverse of **electrolysis**. Gases such as hydrogen (the **fuel**) and oxygen are passed over special porous **electrodes** and a reaction takes place (in this case the gases are converted into water) i.e.

$$2H_2(g) + O_2(g) \rightarrow 2H_2O(l)$$

As this is happening, an electric current flows round the circuit. Fuel cells have been used in spacecraft, but commercial versions have not yet been made for use on Earth.

fullerenes *See* **carbon**.

functional group (in organic chemistry) The atom or group of atoms present in a molecule that is responsible for the characteristic properties of that molecule. Some examples of functional groups are:

$-CO_2H$ carboxylic acid
$-CHO$ aldehyde
$-NH_2$ amine

The functional groups of organic chemistry correspond to the **radicals of inorganic chemistry**.

fungicide *See* **biocide**.

fusion **1.** Another term for **melting**. *See also* **latent heat**.

2. The coming together, in some **nuclear reactions**, of two atoms to form a single atom. This is the opposite of **fission** and can involve the release of enormous quantities of energy. Hydrogen bombs are fusion weapons. Attempts are being made to use fusion reactions to generate electricity.

G

g   The **state symbol** used to denote that a substance is a gas, for example, $CO_2(g)$, $O_2(g)$.

galena   A mineral form of lead(II) sulphide (PbS) which is the principle ore of lead.

galvanizing   A process for coating iron and steel sheeting with a thin layer of zinc. This is called *galvanized iron*. Zinc is more resistant to corrosion than iron and so it can protect the metal. Also, if the coating is scratched and the iron and zinc come into contact with a liquid and an electrochemical **cell** is set up, the zinc reacts rather than the iron – it is more reactive. Thus galvanizing protects the iron even when the protective layer is broken. *See also* **sacrificial anode**.

gamma rays (γ)   A form of high energy electromagnetic radiation. Gamma rays are produced in **nuclear reactions** and have great penetrating powers. They are similar in nature to **X-rays** but have greater energy. Gamma rays are used in the treatment of cancer – they kill body cells. They are also used to sterilize substances such as surgical instruments and animal foodstuffs.

gas   A **state of matter** in which atoms and molecules have few bonds between them and consequently have a large amount of freedom of movement. Gas particles move at high velocity and in random directions.
   When heat energy is supplied to a liquid, the atoms or molecules are given increased **kinetic energy**; this may be sufficient to overcome the bonds that hold them together in the liquid state. If this happens the liquid boils and turns into a gas or **vapour**. In a gas, atoms or molecules fill the whole container and their collisions with the walls of the container exert a pressure.

gas laws   The laws that describe the behaviour of gases. The two main ones are **Boyle's law** and **Charles' law**. If they are combined, it is clear that for a fixed mass of gas:

$$\frac{\text{Pressure} \times \text{Volume}}{\text{Temperature}} = \text{constant}$$

or:

$$\frac{PV}{T} = \text{constant} \ (T \text{ is in kelvin})$$

This means that if gases are compared under different conditions, e.g. different temperature and pressure then:

$$\frac{P_1 V_1}{T_1} = \frac{P_2 V_2}{T_2} = \text{constant}$$

condition 1   condition 2

This allows us to obtain much useful information about gases by calculation rather than by measurement. *See also* **STP**.

gel   A **colloidal dispersion** that has set to form a jelly. Examples are the fruit jellies which you can make at home, and the light-sensitive layer on photographic film. These are both based on **gelatin**.

gelatin   A complex mixture of **amino acids** that is produced by boiling animal bones, hides and cartilage in dilute acids. Gelatins are widely used in foods, glues and photography. *See also* **gel**.

general formula   (in **organic chemistry**) A formula that shows the relative numbers of the different atoms in terms of the variable n for all the members of a particular family of compounds. The actual formula of a particular compound is found by substituting for n. For example, the general formula of the **alkanes** is $C_n H_{2n+2}$, therefore the formulae for its compounds can be deduced as follows:

Methane has 1 carbon $\therefore$ $n = 1$ $\therefore$ formula $= CH_4$

Butane has 4 carbons $\therefore$ $n = 4$ $\therefore$ formula $= C_4 H_{10}$

geochemical cycle   *See* **rock cycle**.

germicide   A substance that is used to reduce or prevent the growth of microorganisms. Many modern germicidal compounds are based on the molecule **phenol**. Antiseptics and disinfectants are types of germicide.

giant structure   A chemical structure in which large numbers of the constituent particles (atoms or ions) are arranged in a crystal **lattice**. Each particle has a strong force of attraction for all the other particles that are near to it. In this way attractive forces are spread through the structure and giant structures tend to have high melting and boiling points. Ionic substances have giant structures as do most elements, including all solid metals and several non-metals.

glass   A hard and usually transparent mixture of silicates. It has an **amorphous** (non-crystalline) structure. Glass is strong but also **brittle**.

The cheapest and commonest kind of glass is called *soda glass*. It is made by heating together **sand**, sodium carbonate, calcium oxide and broken glass (*cullet*). The hot liquid is cooled to below its freezing point very rapidly so that the glass does not have time to form crystals. It is therefore known as a *supercooled liquid*. It is because it is non-crystalline that it is transparent.

In *lead glass*, the sodium carbonate is replaced by lead(II) oxide. This gives a glass with a high refractive index. It is used in making **crystal glassware**. *See also* **borosilicate glass, silica**.

global warming The increase in the temperature of the **Earth's atmosphere**. Since 1860, the temperature is estimated to have risen by 0.55 °C, and the rate of increase is thought to be getting larger. The increasing amount of **greenhouse gases** in the atmosphere is thought to be the cause, although this is still controversial.

glucose A **monosaccharide** molecule. It is found in plants and honey. All the **sugar** and **starch** that enters our bodies is converted to glucose. It is then used to provide energy.

gluten A mixture of **proteins** found in wheat. Gluten is an elastic material which stretches as bread rises. The heat of the oven sets the gluten into a 3-D structure which gives bread its shape and texture.

glycol A material consisting of ethane-1,2-diol which is used as **antifreeze** material in engines. The compound contains two **alcohol** (OH) groups and is made from **ethene** by **oxidation** to epoxyethane and then water is added to give the diol. Ethene is obtained by the **cracking** of petroleum.

glycol *Oxidizing ethene to obtain glycol.*

gold A valuable metal that is prized for its use in jewellery. It is found uncombined with other elements and the major deposits are in South Africa and the Russian Federation. Chemically, gold is very **inert**, reacting only with vigorous **oxidizing agents** such as chlorine and certain acids, e.g. *aqua regia*. Gold is a soft metal and for most uses it is alloyed with **copper** or **silver**.

Pure gold is known as 24-carat gold. 9-carat gold is 9 parts gold to 15 parts copper, i.e. 37.5% gold. This is hard and is commonly used in jewellery.

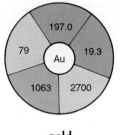

gold

gram (g)   A unit of **mass**. One gram is $\frac{1}{1000}$ of a kilogram and is used in all scientific work. The symbol 'g' is used, e.g. 100 g.

granite   An **igneous** rock widely used as a building material.

graphite   An **allotrope** of carbon. It is found naturally as *plumbago*. **Charcoal** consists of small particles of graphite. Graphite is a **giant structure** with the carbon atoms bonded together in a hexagonal arrangement in sheets or planes.

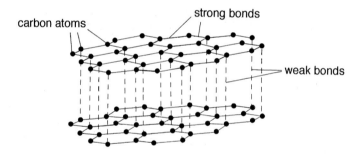

graphite *The giant structure of bonded carbon atoms. Note the strong bonds between atoms in each plate and the weak bonds between the atoms in the different plates.*

There are strong **bonds** between the atoms in the planes but the bonds between the atoms in different planes are weak. It is possible for the planes to slip over each other.

Because of this property, graphite is very useful as a lubricant and, of course, in pencils where layers of graphite are left behind on the paper.

Graphite also finds use as *electrodes* in the extraction of elements, such as sodium, aluminium and chlorine, since it is an electrical **conductor**.

greenhouse effect   The increase in global temperature caused by the build-up of gases in the **Earth's atmosphere** which absorb the Sun's energy that has been re-radiated by the Earth. This trapped energy keeps the planet 40 °C warmer than it would otherwise be. In doing this it is rather like a greenhouse – hence the term – but the mechanism for warming the Earth and keeping a greenhouse hot are quite different. *See also* **global warming, greenhouse gas**.

greenhouse gas A gas whose molecules contribute to the **greenhouse effect** and **global warming** by absorbing energy emitted or radiated from the surface of the Earth. Water vapour is the gas with the strongest greenhouse effect. Other gases (in decreasing power) are carbon dioxide, methane, **CFC** gases, nitrous oxide and **ozone**. Various steps are being taken to reduce the emission of greenhouse gases. Increased tree planting is also taking place as trees absorb carbon dioxide from the atmosphere. There is much debate over how greenhouse gases should be reduced and over what timescale.

group A column within the **periodic table,** containing elements with similar chemical properties. Each group has a characteristic **electronic configuration**: the outermost electron shell contains the same number of electrons for each member of the group. This is the same as the group number.

Group	Elements
I	lithium, sodium, potassium
II	beryllium, magnesium, calcium
III	boron, aluminium, gallium
IV	carbon, silicon, germanium
V	nitrogen, phosphorus, arsenic
VI	oxygen, sulphur, selenium
VII	fluorine, chlorine, bromine, iodine
0	helium, neon, argon, xenon, radon

group *Examples of the elements contained in each group.*

gypsum A **mineral** form of calcium sulphate ($CaSO_4.2H_2O$).

H

Haber process A process for producing **ammonia** from hydrogen and nitrogen. Nitrogen is obtained from the air and hydrogen from the **steam reformation** of natural gas. Any sulphur is removed from the raw materials by **desulphurization** and steam is then added:

$$CH_4(g) + 2H_2O(g) \rightarrow CO_2(g) + 4H_2(g)$$

The carbon dioxide is removed by passing the gas through potassium carbonate solution:

$$CO_2(g) + K_2CO_3(aq) + H_2O(l) \rightarrow 2KHCO_3(aq)$$

Air is then added to give a $3:1$ hydrogen : nitrogen mixture. The gases are reacted together at $500\,°C$ and a pressure of 200 atmospheres in the presence of an iron **catalyst**:

$$N_2(g) + 3H_2(g) \rightleftharpoons 2NH_3(g)$$

This is an **equilibrium** reaction and the conditions are chosen in order to produce as much ammonia as possible in the shortest time. Under these conditions about 15% of the reactants are converted to ammonia. Using a lower temperature will generate more ammonia but at a much slower **rate**. Conversely, using a higher temperature will produce the ammonia more quickly, but the **yield** will be lower. The chosen conditions are the optimum ones.

haematite A mineral form of iron(III) oxide (Fe_2O_3) which is one of the main raw materials for the production of **iron**.

haemoglobin The red pigment that is present in red blood corpuscles. It contains an atom of iron in the complex molecule. Oxygen reacts with haemoglobin to form *oxyhaemoglobin*, and is then released from the complex in the parts of the body where it is needed.

Carbon monoxide also reacts with haemoglobin, forming *carboxyhaemoglobin*. This is a very stable compound and prevents the carriage of oxygen. This is why carbon monoxide is such a dangerous **poison**.

It is important that the body obtains sufficient iron each day to maintain the correct levels of haemoglobin. The daily requirement is about 11 mg. Insufficiency leads to the condition known as **anaemia**.

half-life The time taken for one half of the radioactive **isotopes** originally present in a sample to decay. When a radioactive isotope gives off **alpha particles** or **beta particles** (decays) it changes into a different isotope, so as time goes on there is increasingly fewer of the original isotopes. The decay of an isotope is usually traced by measuring the rate at which particles are emitted. This rate is proportional to the number of original nuclei present.

Isotope	Half-life
Carbon-14	5730 years
Oxygen-20	14 seconds
Copper-64	756 minutes
Uranium-234	250,000 years

half-life *The half-lives of different isotopes.*

The graph shows the decay of an isotope with a half-life of one minute, i.e. in each minute the number of particles emitted falls by half e.g. 4000–2000, 2000–1000, etc.

There is great variety in the length of half-lives fromisotope to isotope. The half-life of any one isotope is constant, however, under all conditions of temperature and pressure.

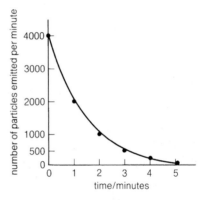

half-life *The decay of an isotope with a half-life of one minute.*

half-reaction A way of representing a reaction by 'splitting' the process into two halves and considering each half separately. For example, the **displacement** of copper ions from solution by zinc:

$$Cu^{2+}(aq) + Zn(s) \rightarrow Zn^{2+}(aq) + Cu(s)$$

can be viewed as:

$$\text{(a) } Cu^{2+}(aq) + 2e^- \rightarrow Cu(s)$$

$$\text{(b) } Zn(s) \rightarrow Zn^{2+}(aq) + 2e^-$$

From these half-reactions it is easy to see that the zinc is being oxidized by the copper ions (*see* **oxidation**) and that the copper ions are being reduced by the zinc (*see* **reduction**). Half-reactions are particularly useful for studying **redox** reactions.

halide   A compound of a **halogen** with another **element**. Examples include sodium bromide (NaBr) and calcium fluoride (CaF_2). Metal halides tend to be **ionic**, whereas nonmetal halides have **covalent bonds**. Metal halides can be produced by reacting the elements together:

$$2Na(s) + Cl_2(g) \rightarrow 2NaCl(s)$$

Hall–Hérault cell   An electrolytic **cell** developed in 1886 for the production of aluminium.

halogens   The elements in group VII of the **periodic table**. They are all poisonous, non-metallic elements. At room temperature, **fluorine** and

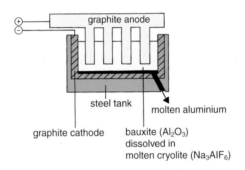

Hall–Hérault cell

chlorine are gases, **bromine** is a liquid and **iodine** is a solid. The halogens are all **oxidizing agents**. Their oxidizing power and chemical reactivity decreases in the order $F_2 > Cl_2 > Br_2 > I_2$, i.e. down the the group.

Halogens react vigorously with metals and hydrogen forming **halides**. They all contain seven electrons in the outer shell of the atom and form univalent **anions**, e.g. Cl^-, Br^-.

hardening   A process in which unsaturated oils (*see* **unsaturated compounds**) are reacted with hydrogen to produce **saturated compounds**, which are more suitable for the production of vegetable oils and margarines.

hard water   Water that contains dissolved minerals such as **calcium** and **magnesium** salts. These dissolve in the water when it passes over and through porous rocks such as **limestone, chalk** and **gypsum**).

Soft water, in contrast, collects in areas where the rocks are insoluble in water, such as the granite areas of the Lake District, Cornwall and the Highlands of Scotland. Soft water lathers relatively easily.

Hard water does not easily lather with **soap** but forms a scum, caused by the metal ions that react chemically with the soap to form insoluble calcium salts. Soapless **detergents** do not do this.

There are two types of hardness: *temporary hardness* and *permanent hardness*. Temporarily hard water contains dissolved calcium hydrogencarbonate as a result of slightly acidic rain water dissolving chalk and limestone.

$$CaCO_3(s) \; + \; H_2O(l) \; + \; CO_2\,(g) \; \rightleftharpoons \; Ca(HCO_3)_2(aq)$$

| calcium | rainwater | calcium |
| carbonate | | hydrogencarbonate |

The hardness is called temporary since it is simply removed by boiling, which reverses the reaction to form the insoluble carbonate, which is precipitated. The use of kettles and immersion heaters with hard water becomes increasingly difficult and expensive with time because the build-up of calcium carbonate on the heating elements makes them progressively more inefficient.

Permanently hard water contains dissolved calcium sulphate and magnesium sulphate. These are not removed by boiling. They can be removed by adding sodium carbonate to precipitate the insoluble carbonates.

A wide range of products are used to make water softer through the use of **ion exchange**.

hazard warnings *See Appendix B.*

heat energy **Energy** derived from the movement of atoms and molecules, i.e. their **kinetic energy**. All substances possess heat energy. The higher their **temperature**, the more energy they possess. Heat energy is taken in during **endothermic reactions** and given out during **exothermic reactions**.

heat exchange A process in which **heat energy** contained in one material is transferred to another. In industry, this may take place in *heat exchangers* and heat produced in one part of a process can be transferred to another part, thus saving energy and money. Chemical production plants are carefully designed in order to make heat exchange as efficient as possible. Heat exchange also takes place in the home. Central heating systems, freezers and fridges all use this process.

heat of combustion The **heat energy** released when one **mole** of a substance is burnt in oxygen (with no change in volume).

heat of neutralization   The **heat energy** released when one **mole** of hydrogen ions and one mole of hydroxide ions react to form water.

heat of reaction   The **heat energy** taken in when one **mole** of a substance is formed from its elements.

heavy metal   Any metal whose density is more than 5 $g\,cm^{-3}$. Examples include lead and mercury.

helium (He)   A **noble gas**. Helium is found in natural gas (up to 6%) and is present in the atmosphere to a very small extent. It is completely unreactive and the gas is monatomic (*see* **monatomic molecule**).
  Helium is used in airships as it is eight times less dense than air and non-flammable. Helium mixed with oxygen is also used in breathing apparatus for deep-sea divers.

heterogeneous reaction   A chemical reaction that takes place between substances that are in different physical **states**, i.e. gases, liquids and/or solids. *Hetero-* means different. Examples of heterogenous reactions include:

$$Zn(s) + 2HCl(aq) \rightarrow ZnCl_2(aq) + H_2(g)$$

$$2H_2O(l) + 2Na(s) \rightarrow 2NaOH(aq) + H_2(g)$$

$$2Na(s) + Cl_2(g) \rightarrow 2NaCl(s)$$

See also **homogeneous reaction**.

homogeneous reaction   A chemical reaction in which all the reactants and products are in the same physical **state**. *Homo-* means same. Examples of homogenous reactions include:

$$Fe(s) + S(s) \rightarrow FeS(s)$$

$$N_2(g) + 3H_2(g) \rightarrow 2NH_3(g)$$

homologous series   A series of compounds with the same **general formula**. For example:
(a) **alkanes** C_nH_{2n+2}
(b) **alkenes** C_nH_n.

The compounds have the same **functional groups** and similar chemical properties. The physical properties of the series of compounds show a steady change along the series.

Alkanes: C_nH_{2n+2}

$$\begin{array}{c}
H \\
| \\
H-C-H \\
| \\
H
\end{array}$$
methane CH_4
$n = 1$

$$\begin{array}{c}
H \quad H \\
| \quad | \\
H-C-C-H \\
| \quad | \\
H \quad H
\end{array}$$
ethane C_2H_6
$n = 2$

$$\begin{array}{c}
H \quad H \quad H \\
| \quad | \quad | \\
H-C-C-C-H \\
| \quad | \quad | \\
H \quad H \quad H
\end{array}$$
propane C_3H_8
$n = 3$

homologous series

hydrate A compound that has water chemically combined within it. *Hydrated* **salts**, i.e. those which contain **water of crystallization** are the best examples, such as hydrated copper(II) sulphate, $CuSO_4.5H_2O$ and hydrated calcium chloride, $CaCl_2.6H_2O$.

hydration A chemical reaction where a compound either reacts with a molecule of water to form a new substance or reacts with water to form a hydrate that contains **water of crystallization**. Examples of such reactions are:

$$C_2H_4(g) + H_2O(g) \rightarrow C_2H_6O(g)$$
ethene   water   ethanol

$$CuSO_4(s) + 5H_2O(l) \rightarrow CuSO_4.5H_2O(s)$$
anhydrous   hydrated
copper(II) sulphate   copper(II) sulphate

hydride Any compound that contains hydrogen and one other element only. Examples are ammonia, NH_3; water H_2O; and hydrogen sulphide, H_2S.
  The hydrides of non-metals are covalent compounds. Some metal hydrides contain the ion H^-, for example Na^+H^-. These compounds are very reactive and will readily decompose water.

hydrocarbon A compound of hydrogen and carbon only, for example **alkanes**, **alkenes** and **alkynes** such as methane, CH_4; ethyne C_2H_2; ethene,

C_2H_4 and benzene, C_6H_6. *See also* **cracking**, **petroleum**, **saturated compound**, **unsaturated compound**.

hydrochloric acid (HCl(aq))   A solution of **hydrogen chloride** in water; it contains **chloride** and **oxonium ions**. The maximum **concentration** of the solution is 36% (about 11 mol dm^{-3}). It is a **monobasic acid** and produces **salts** called **chlorides**:

$$Fe(s) + 2HCl(aq) \rightarrow FeCl_2(aq) + H_2(g)$$

$$Mg(s) + 2HCl(aq) \rightarrow MgCl_2(aq) + H_2(g)$$

It is a strong acid (*see* **strength of acids and bases**) being fully dissociated into $Cl^-(aq)$ and $H_3O^+(aq)$ ions in dilute solution. It releases carbon dioxide from **carbonates** and **hydrogencarbonates** and can be oxidized to **chlorine**:

$$MnO_2(s) + 4HCl(aq) \rightarrow MnCl_2(aq) + 2H_2O(l) + Cl_2(g)$$

hydrogen (H)   A gaseous diatomic element (*see* **diatomic molecule**). A hydrogen atom consists of one proton and one electron. The isotope **deuterium** contains a single neutron in the nucleus of the atom. There is a further isotope called *tritium* which has two neutrons in its nucleus and is radioactive.

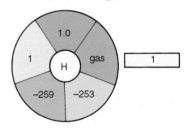

hydrogen

    Hydrogen is very reactive. It can form **covalent bonds** by sharing electrons. For example:

$$2H_2(g) + O_2(g) \rightarrow 2H_2O(g)$$

$$C_2H_4(g) + H_2(g) \rightarrow C_2H_6(g) \text{ (}hydrogenation\text{)}$$

Hydrogen ions can be produced by losing the electron to form H^+ or by gaining an electron to form H^- (present in some **hydrides**). The positive hydrogen ion (H^+) is such a small, reactive species (it is a lone proton) that it cannot exist alone in solution. In **aqueous solutions** it reacts with the water to form the **oxonium ion**. Acids contain the oxonium ion.

    Hydrogen is a **reducing agent**. It is used to make **methanol** and **nylon** and is growing in importance as a fuel. It is used to convert vegetable oils into margarine. Large quantities are used in the **Haber process**.

    Industrially, the gas is made from petroleum by the **steam reformation** of **naphtha** and natural gas. In the laboratory, the gas is made by reacting a metal with an acid other than nitric acid, e.g.:

$$Zn(s) + 2HCl(aq) \rightarrow ZnCl_2(aq) + H_2(g)$$

It is also released by the electrolysis of aqueous solutions which contain ions of elements above hydrogen in the electrochemical series, e.g. NaCl(aq), $Mg(NO_3)_2$(aq), and by the reaction of water with alkali and alkaline earth metals, e.g. sodium, calcium.

Hydrogen is a flammable gas that forms explosive mixtures with oxygen. Great care must be taken in its preparation, collection and use.

hydrogen bond A weak chemical **bond** formed between hydrogen atoms and atoms of oxygen or nitrogen. Although weaker than **covalent bonds** or **ionic bonds**, hydrogen bonds affect the physical properties of compounds. Molecules in water and ice are extensively hydrogen-bonded to each other.

hydrogen bond *Molecules in water and ice.*

hydrogenation A process in which an **unsaturated compound** is turned into a **saturated compound** by the addition of hydrogen. A **catalyst** such as nickel is used. For example:

ethene   ethane

This process is sometimes known as **hardening**, and is widely used in the food industry, e.g. in the production of margarines from vegetable oils.

hydrogencarbonates (HCO_3^-) The **acid salts** of *carbonic acid* H_2CO_3. The best known examples are sodium hydrogencarbonate, $NaHCO_3$ and calcium hydrogencarbonate, $Ca(HCO_3)_2$.

Sodium hydrogencarbonate is the familiar household chemical *bicarbonate of soda*. It is found in self-raising flour and **baking powder**.

Calcium hydrogencarbonate is found in **hard water**. The compound dissolves in the water when it passes over **carbonate** rocks such as **limestone**:

$$CaCO_3(s) + CO_2(aq) + H_2O(l) \rightarrow Ca(HCO_3)_2(aq)$$

hydrogen chloride (HCl) A colourless gas that is very soluble in water. An **aqueous solution** is called **hydrochloric acid**. The gas can be made by reacting together the elements or by treating sodium chloride with concentrated sulphuric acid:

$$H_2(g) + Cl_2(g) \rightarrow 2HCl(g)$$

$$NaCl(s) + H_2SO_4(l) \rightarrow NaHSO_4(s) + HCl(g)$$

Hydrogen chloride gas reacts with ammonia to form dense white fumes of ammonium chloride:

$$NH_3(g) + HCl(g) \rightarrow NH_4Cl(s)$$

The bonding in hydrogen chloride is covalent but ions are formed when it dissolves in a **polar solvent** such as water.

hydrogen halides Gaseous compounds formed between hydrogen and the **halogens**. They are **covalent compounds** that readily dissolve in water to form an acidic solution that contains the **halide** ion, e.g.

$$HBr(g) + H_2O(l) \rightarrow H_3O^+(aq) + Br^-(aq)$$

Hydrogen fluoride	HF
Hydrogen chloride	HCl
Hydrogen bromide	HBr
Hydrogen iodide	HI

hydrogen ion A **proton**, formed by the loss of an electron from the hydrogen atom:

$$H \rightarrow H^+ + e^-$$

This is a very small and very reactive particle. In solution, it is chemically combined with the solvent molecules. In water this is represented by the **oxonium ion**:

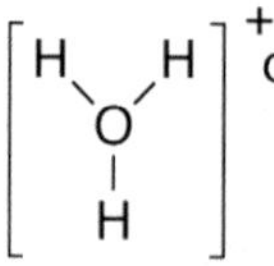

The concentration of hydrogen ions in an **aqueous solution** is expressed in **pH** units, and gives a measure of the degree of acidity of the solution. *See also* **hydroxide ion**.

hydrogen peroxide (H_2O_2) A compound that is usually used as an **aqueous solution**. It readily decomposes to give oxygen:

$$2H_2O_2(aq) \rightarrow 2H_2O(l) + O_2(g)$$

A manganese(IV) oxide **catalyst** speeds up the reaction. Hydrogen peroxide is used as a **disinfectant** and **bleach** in the home and is a powerful **oxidizing agent**. It bleaches hair to a blonde colour, and is used in industry for bleaching paper pulp and natural fibres. It is less destructive than the more powerful bleach chlorine.

hydrogen sulphide (H_2S) A colourless gas with a sweetish, sickly odour similar to that of rotten eggs. The gas is produced when organic matter containing sulphur rots. It is often found associated with petroleum. It is made in the laboratory by the action of an acid on a metal sulphide, e.g.:

$$FeS(s) + 2HCl(aq) \rightarrow FeCl_2(aq) + H_2S(g)$$

Hydrogen sulphide gas is extremely poisonous. It can be detected by the fact that it turns a piece of filter paper, soaked in a lead(II) salt, black. This is because insoluble lead(II) sulphide is formed:

$$Pb^{2+}(aq) + H_2S(g) \rightarrow PbS(s) + 2H^+(aq)$$
$$\text{black}$$

A solution of hydrogen sulphide in water is a **weak acid** and the gas is a **reducing agent**.

hydrolysis The process of **decomposition** of a substance by the action of water. The water is also decomposed. **Esters** may be hydrolysed, as the example shows:

ethyl ethanoate   water   ethanoic acid   ethanol

In general, a catalyst of hydrogen or hydroxide ions is needed for hydrolysis to occur.

hydrophilic Water-loving. The term is used to describe parts of molecules that readily dissolve in water; for example, the ionic end of a **detergent** molecule is hydrophilic. The other end of the molecule is water-hating or *hydrophobic*. The hydrophobic part of a molecule does not dissolve in water and is not attracted to water.

Detergents work because the hydrophilic end of the molecule is attracted to water, while the hydrophobic end bonds with grease or oils in the material being cleaned. *See also* **soap**.

hydrophobic *See* **hydrophilic**.

hydroxide ion (OH⁻) An **ion** found in all **alkalis**, e.g. sodium hydroxide, and in **alkaline** solutions. It is present in all **aqueous solutions** because of the **dissociation** of water:

$$H_2O \rightleftharpoons H^+ + OH^-$$

Solutions with more hydroxide ions than hydrogen ions are described as alkaline and have a **pH** greater than 7.

Group I hydroxides are soluble in water. Some other hydroxides are sparingly soluble, e.g. $Mg(OH)_2$, $Ca(OH)_2$. Insoluble hydroxides can be precipitated by the use of a soluble hydroxide. For example:

$$Pb(NO_3)_2(aq) + 2NaOH(aq) \rightarrow Pb(OH)_2(s) + 2NaNO_3(aq)$$

hygroscopic (used of a substance) Able to take in up to 70% of its own mass of water without dissolving or getting wet. Examples are copper(II) oxide, sodium chloride and **silica gel**. *See also* **deliquescent substances**.

I

ice Solid, crystalline water. Its melting point at a pressure of 1 atmosphere is 0 °C. The regular, crystalline structure means that frozen water can take on the patterns that we see in snowflakes. Ice is less dense than water and it floats. *See also* **crystal**.

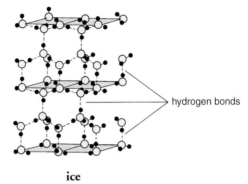

hydrogen bonds

ice

igneous rocks Rocks formed when molten rock (*magma*) cools and solidifies. Igneous rocks are crystalline and they normally contain minerals. They make up 90% of the Earth's crust. Examples are granite, gabbro and basalt. *See also* **crystallization**, **rock cycle**.

ignition temperature The minimum temperature at which a substance will burn in air.

immiscible Not capable of being mixed. When two liquids do not mix together but form two layers, one liquid on top of the other, they are described as being immiscible.

In an immiscible mixture, one liquid will probably be **polar**, e.g. water, and the other non-polar, e.g. ether. Two polar liquids will mix completely, e.g. water and ethanol, as will two non-polar liquids, e.g. ether and tetrachloromethane. Two immiscible liquids may be separated using a separating funnel. *See also* **miscible**.

indicator A substance that changes colour under different conditions, e.g. **litmus** is red in **acid** solution and blue in **alkaline**. Indicators such as litmus that are a different colour depending on the **pH** are useful in acid-base **titrations** where they can be used to indicate the **end point** of the reaction. They are also used in titrations between metal ions and **complex ions**. *See also* **universal indicator**.

inert Unreactive. *See* **CFCs**, **noble gases**.

inert gases *See* **noble gases**.

infrared (IR) radiation Invisible electromagnetic radiation with an energy slightly lower than that of visible radiation. It is produced by warm objects. Our skins are sensitive to this radiation, which we experience as heat. Photographic film can be made that is sensitive to infrared radiation and can be used where there is not enough visible light to take normal photographs.

inhibitor A substance that will slow down or stop a chemical reaction.

initiator A substance that is used to start off a chemical reaction such as a **polymerization** or an explosion.

inorganic (used of substances) Describing substances that contain no carbon or that are carbides or carbonates, or oxides or sulphides of carbon. Such compounds are generally obtained from the Earth's crust or atmosphere. *See also* **organic**.

inorganic chemistry The study of the elements and **inorganic** compounds. It is concerned with non-organic aspects of chemistry and excludes all **organic** compounds, e.g. **alcohols**, **esters**, **ethers**, **hydrocarbons**, etc. *See also* **organic chemistry**.

insecticide *See* **biocide**.

insoluble (used of a substance) Not able to **dissolve** in a **solvent**.

insulator A substance that is a poor **conductor** of either **heat** or electricity.
  Non-metallic elements are usually insulators, as are most solid compounds and **polymers**. **Graphite** is an exception. Examples of efficient insulators are:
• expanded polystyrene;
• rubber;
• mineral wools, e.g. Rockwool;
• glass fibre.

iodides Compounds of iodine and another element. Examples are potassium iodide, KI, and hydrogen iodide, HI.

iodine (I) A shiny, grey non-metallic element. It is a **halogen** and its structure is that of a **diatomic molecule**. Iodine is extracted mainly from sodium iodate(V) ($NaIO_3$), and to a lesser degree from seaweed. It is used to produce animal feeds, **catalysts**, printing inks, **dyes** and pharmaceuticals.

Silver iodide is used in photographic emulsions.
The human body needs 0.07 mg of iodine per
day. Its principal use in the body is in the
production of the hormone *thyroxine*.
Potassium iodide is often added to table salt
to provide this iodine.

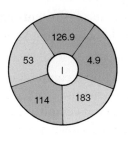

iodine

Iodine will react with some metals directly to
form **iodides**. It will react with hydrogen and
chlorine forming HI and ICl. The vapour of
the element is purple, choking and **caustic**.

ion An **atom** or group of atoms that possess an electrical charge. When
an atom gains or loses an electron it becomes an ion. **Cations** have a
positive charge, **anions** have a negative charge, e.g. sodium Na^+, oxide O^{2-}.
Atoms tend to gain or lose electrons to produce an ion with a noble gas
configuration (*see* **noble gas structure**). Groups of atoms (radicals) can also
possess a charge. For example, the sulphate group, SO_4^{2-}, nitrate group,
NO_3^-, ammonium group, NH_4^+.

ion exchange The exchange of ions of the same electrical charge between
a solution and a solid in contact with it. It is possible to purify water using
ion exchange by passing it over a resin (called an *ion exchange resin*) in a
tube. With seawater the sodium ions would be replaced by hydrogen ions
and the chloride ions by hydroxide ions. In other words, sodium chloride
would be exchanged for water. The ions are bonded to the resin and
exchange occurs. The product is *de-ionized* water. This method can be used
for softening water, e.g. calcium ions are replaced by sodium ions.

ionic Describing a material that contains **ions** or that is held together by
ionic bonds.

ionic bonds Chemical **bonds** that are formed as a result of the electrostatic attractive forces between negatively and positively charged **ions**.
Ionic bonds occur in compounds of non-metals from groups VI and VII
and metallic elements, e.g. Na^+Cl^-, and also in compounds involving
radicals such as sulphate and nitrate, for example $Cu^{2+}SO_4^{2-}$ and $K^+NO_3^-$.
Ionic compounds have **giant structures**.

ionic equation A chemical **equation** that shows only those **ions** that
take part in a chemical reaction. For example, the reaction between
sodium hydroxide and hydrochloric acid can be described by the molecular
equation:

$$NaOH(aq) + HCl(aq) \rightarrow NaCl(aq) + H_2O(l)$$

This reaction is an ionic one and can be represented in an ionic form:

$$Na^+OH^-(aq) + H^+Cl^-(aq) \rightarrow$$
$$Na^+Cl^-(aq) + H_2O(l)$$

Because Na^+ and Cl^- appears on both sides of the equation they do not take any part in the reaction and can be omitted. The ionic equation is therefore:

$$OH^-(aq) + H^+(aq) \rightarrow H_2O(l)$$

The other ions are termed **spectator ions.**

ionic lattice *See* **giant structure**.

ionization The process by which an **atom** loses or gains **electrons** and becomes an **ion**.

ionizing radiation Radiation that creates **ions** when it passes through living tissue. There are different types of ionizing radiation, including high-energy particles, such as **electrons** or **protons**, and **ultraviolet radiation, X-rays** and **gamma rays**. Each type penetrates materials to a different extent. Some types of ionizing radiation is used in medicine for both diagnosis and treatment. *See also* **radioactivity**.

iron The most widely used metallic element. It is mainly encountered in **steel** alloys and in this form it is used for building girders, machine bodies (cars, cookers, fridges), containers (boxes, cans, drums), tools, utensils and many other items of everyday life. The metal is extracted from ores such as haematite (Fe_2O_3) and magnetite (Fe_3O_4) by reduction with carbon monoxide in the **blast furnace**. Iron produced in this way is brittle and is usually made into many different kinds of steel to strengthen it and to give it special properties.

One of the main problems with iron is that it rusts, i.e. it oxidizes in air to produce a soft, crumbly oxide.

Iron — (moist air) → Iron(III) oxide (Fe_2O_3)
(strong, useful metal)   (weak, worthless oxide)

Iron is a **transition metal** and can have **valency** of 2 or 3. The metal reacts with dilute acids to form iron(II) compounds, e.g.:

$$Fe(s) + 2HCl(aq) \rightarrow FeCl_2(aq) + H_2(g)$$

but will give iron(III) compounds when reacted
with vigorous **oxidizing agents** such as chlorine:

$$2Fe(s) + 3Cl_2(g) \rightarrow 2FeCl_3(s)$$

Iron has a **reversible reaction** with steam:

$$3Fe(s) + 4H_2O(g) \rightleftharpoons Fe_3O_4(s) + 4H_2(g)$$

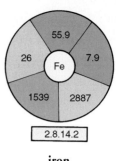

2.8.14.2

iron

Iron is essential for human health, particularly
for the transport of oxygen round the body in
the blood. *See also* **anaemia**, **steel manufacture**.

iron compounds Iron forms compounds with **valency** 2 and 3. Iron(II)
compounds tend to be green in colour whilst iron(III) compounds are
yellow or brown. Iron(III) compounds can be identified by their giving a
red colour with potassium thiocyanate solution (KCNS). Iron(II)
compounds do not do this but do produce a blue colour with a **solution** of
potassium hexacyanoferrate(III) $K_3Fe(CN)_6$.
 Furthermore, when reacted with sodium hydroxide solution, iron(II)
solutions produce a muddy green **precipitate**, whereas iron(III) solutions
produce a rust-brown precipitate.

Iron(III) oxide Fe_2O_3 (haematite)	A yellowish/brown pigment used in the paper, linoleum and ceramics industries. It is used as a catalyst and as a polish, e.g. in jeweller's rouge.
Iron(III) chloride $FeCl_3$	Produced by reacting the elements together. It is an important catalyst and also finds use in the purification of water and in the production of pharmaceutical products.
Iron(II) sulphate $FeSO_4.7H_2O$	Iron(II) sulphate can be recovered from the waste materials left behind in the electroplating processes. It is an important compound, used in the preservation of wood, in inks and inlithography.

isomers Two or more different compounds with the same **molecular
formula**. Because the compounds are different, they have different
properties. For example, ethanol and methoxymethane are isomers, both
having the molecular formula C_2OH_6.

isotopes Atoms of the same element containing different numbers of
neutrons (different **mass number**). Atoms of an element always contain the
same number of **protons** (same **atomic number**). A sample of chlorine

contains atoms which have 18 neutrons (76% of the total) and atoms which have 20 neutrons (24%). Thus, chlorine is said to have two isotopes:

$$^{35}_{17}\text{Cl} \qquad ^{37}_{17}\text{Cl}$$

The ratio of the isotopes in a given sample of an element is always constant. Most elements are found in isotopic forms. Examples include:

Bromine	$^{79}_{35}\text{Br}$ (51%)	$^{81}_{35}\text{Br}$ (49%)
Carbon	$^{12}_{6}\text{C}$ (99%)	$^{13}_{6}\text{C}$ (1%)
Magnesium	$^{24}_{12}\text{Mg}$ (79%)	$^{25}_{12}\text{Mg}$ (10%)
		$^{26}_{12}\text{Mg}$ (11%)

The ones quoted above are all *stable isotopes*. However, many more unstable, radioactive isotopes exist. A few elements have no (naturally occurring) isotopes, i.e. they have no variation in the number of neutrons that are found in the atom. These include fluorine, gold, iodine, manganese, phosphorus, scandium.

The existence of isotopes accounts for the fact that many elements have a **relative atomic mass** that is not a whole number, for example, that of chlorine is 35.5.

J – K

joule The **SI unit** of **energy** and work. It has the **symbol** J. In scientific usage, quantities are more commonly quoted in **kilojoules** (kJ).

Kelvin temperature scale The kelvin is the **SI unit** of **temperature**. The kelvin (K) is the same as one degree Celsius (*see* **Celsius scale of temperature**). Absolute zero is 0 K.

Note that a temperature expressed in kelvin does not include a degree sign (°), thus the **boiling point** of water is simply 373 K.

Kelvin	0	273	310	373	K
Celsius	–273	0	37	100	°C
	absolute zero	freezing point of water	blood heat	boiling point of water	

Kelvin temperature scale *A comparison with the Celsius scale.*

kerosine A product of **petroleum** refining. Its uses include **fuel** for jet aircraft and also household paraffin which can be used for heating and lighting. Kerosines boil between 160 °C and 250 °C.

kilo- A prefix which means one thousand. For example, a kilogram equals 1000 g and one kilometre equals 1000 m.

kilojoule (kJ) One thousand **joules**.

kinetic energy The **energy** a body possesses as a result of its motion. The greater its velocity (speed) the more energy it has. The kinetic energy of a particle is $\frac{1}{2}mv^2$ where m is its **mass** and v its velocity. If the mass is measured in kilograms and the velocity in metres per second the kinetic energy is measured in **joules**.

kinetic theory The theory that all **particles** (**atoms**, **ions** and **molecules**) are constantly moving and that the extent to which movement can occur depends on the temperature. In solids and liquids the extent of movement is restricted by the bonds between adjacent particles. In gases movement is only restricted by the walls of the container. The higher the temperature, the greater the **kinetic energy** the particles possess. *See also* **lattice, change of state**.

krypton (Kr)   A **noble gas**. Traces of the element are found in the Earth's atmosphere. The gas is used in electronic valves and fluorescent tubes. Like all noble gases, krypton is monatomic (*see* **monatomic molecule**).

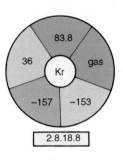

krypton

L

1. A **state symbol** denoting a **liquid**, e.g. $H_2O(l)$, $Hg(l)$, $H_2SO_4(l)$.
2. The unit of volume (litre) which is equivalent to 1 dm^3 or 1000 ml.

L The symbol for the **Avogadro constant**. It is the number of particles in one **mole** of a substance. Its value is 6×10^{23} particles.

lactose   A **sugar** found in all animal milk. It is a **disaccharide** with the formula $C_{12}H_{22}O_{11}$.

lanthanide element   *See* **periodic table**.

large molecule   *See* **macromolecule**.

latent heat   The amount of **heat energy** released or absorbed in a **change of state** at a fixed temperature (e.g. melting point and boiling point). The latent heat of **fusion** is that energy needed to turn one **mole** of solid into a liquid at its melting point.
  Similarly, the latent heat of vaporization is that energy required to turn one mole of liquid into a **gas** at its boiling point. These symbols can be used:

$$\Delta H_m, \text{ where m = melting (fusion)}$$

$$\Delta H_b, \text{ where b = boiling (vaporization)}$$

For example:

Water $\Delta H_m = 6 \text{ kJ/mole}$

$\Delta H_b, = 41 \text{ kJ/mole}$

lattice   A regular arrangement of molecules, atoms or ions within a crystalline solid. Lattices contain large numbers of particles that are arranged in very particular ways. *See also* **crystal**.

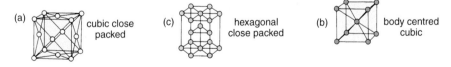

(a) cubic close packed   (c) hexagonal close packed   (b) body centred cubic

lattice *Small sections of three lattices:*
(a) aluminium, (b) sodium, (c) magnesium.

The structure shown overleaf is an example of an ionic lattice:

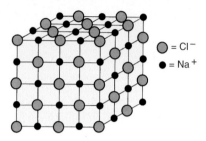

lattice *Ionic lattice.*

law of constant composition   A chemical compound always contains the same elements combined in the same proportions by mass. Examples include the following:
(a) Water always has hydrogen and oxygen within it in the proportion: H:O = 2:16 = 1:8 by mass.
(b) Carbon dioxide always has carbon and oxygen within it in the proportion: C:O = 12:32 = 3:8 by mass.

law of multiple proportions   If two elements, A and B, react together to form more than one compound, the masses of A that combines with a fixed mass of B are in a simple ratio to one another. For example, in water, H_2O, 2 g of hydrogen combines with 16 g of oxygen. In hydrogen, H_2O_2, the same 16 g of oxygen will combine with just 1 g of hydrogen. Thus the masses of hydrogen in the two compounds are in a simple ratio of 2:1.

layer structure   *See* **graphite**.

lead   A dense metallic element that is in group IV of the **periodic table**. It is a soft, **malleable**, grey element that is extracted from galena (PbS). 40% of the lead now produced is recycled from scrap.

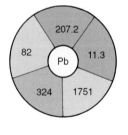

lead

Lead and its **alloys** have wide uses, for example in the lead–acid **accumulator**, in **solders**, and as protection from moisture (around cables and on roofs) and from radioactive isotopes and **X-rays**. Lead was also used to make petrol **additives**, although the use of these has stopped in the UK. Because of its chemical inertness, lead can be used to store sulphuric acid.

lead compounds   Lead forms lead(II) and lead(IV) compounds. Most lead compounds are cumulative **poisons**.

Lead(II) nitrate $Pb(NO_3)_2$	The only common soluble lead compound.
Lead(II) oxide PbO	The yellow oxide, litharge. It is used in the making of glass and enamels.
Lead(IV) oxide PbO_2	This oxide is formed in the lead–acid accumulator.
Tetraethyl lead $Pb(C_2H_5)_4$	The 'antiknock' agent that was added to leaded petrol to make car engines run smoother. It was also a source of pollution from car exhausts.

Le Chatelier's principle   The principle that if a reaction is at **equilibrium** and any of the conditions are changed, further reaction will occur to counter the changes and re-establish equilibrium. For example, in the **Haber process**:

$$N_2(g) + 3H_2(g) \rightleftharpoons 2NH_3(g)$$

If extra nitrogen or hydrogen is added to an equilibrium mixture then more ammonia will be produced and the equilibrium will be re-established. Similarly, if extra ammonia is added, more nitrogen and hydrogen will be produced.

 If the pressure is increased, more ammonia will be produced. This is because the production of more ammonia leads to a lowering of the pressure owing to an overall reduction in the number of gas molecules present. In this way equilibrium is re-established.

 A change of temperature leads to a change in the proportions of the equilibrium mixture. For **exothermic reactions** (such as the Haber process) raising the temperature favours the right to left reaction ($\leftarrow$) (ammonia would react to form more nitrogen and hydrogen). This is because the right to left reaction is an **endothermic reaction**. It therefore counters the rise in temperature by absorbing heat.

Liebig condenser   A glass vessel in which **vapour** is condensed to a liquid during **distillation**.

lime or quicklime   Another name for **calcium oxide**, produced by heating **limestone** above 900 °C. This is an **endothermic reaction**:

$$CaCO_3(s) \rightarrow CaO(s) + CO_2(g)$$

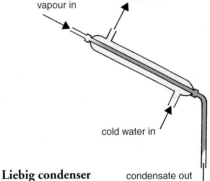

vapour in

water out

cold water in

Liebig condenser

condensate out

Lime is an important industrial chemical that is used in metal production, **refractory** materials, fertilizers, the production of other chemicals and in building materials. It is used to make **slaked lime** and is responsible for turning impurities in the **blast furnace** into **slag**. In the laboratory it is a useful drying agent for ammonia.

limestone   A commonly occurring rock that contains between 50% and 90% calcium carbonate ($CaCO_3$). It is used to make **lime** and **cement** and is used as building stone and hard core for foundations. *See also* **chalk**.

limewater   A dilute solution of the sparingly soluble compound calcium hydroxide. It is an **alkali** and is used to test for carbon dioxide:

$$Ca(OH)_2(aq) + CO_2(g) \rightarrow CaCO_3(s) + H_2O(l)$$

A white **precipitate** is formed and is seen as a milkiness in the solution. If excess carbon dioxide is bubbled through, the precipitate reacts to form the soluble salt, calcium hydrogencarbonate, and the liquid goes clear:

$$CaCO_3(s) + H_2O(l) + CO_2(g) \rightarrow Ca(HCO_3)_2(aq)$$

linear molecules   Molecules that are straight, i.e. their atoms are in a line. All molecules that contain only two atoms must be linear. Other examples include:

carbon dioxide	ethyne	beryllium chloride
O=C=O	H−C≡C−H	Cl−Be−Cl

lipids   *See* **fats**.

liquefaction   The process of turning a **gas** into a **liquid**.

liquid   A **state of matter** in which particles are loosely bonded by intermolecular forces. A liquid always takes up the shape of its container. The particles in the liquid are not fixed in a rigid framework (**lattice**). *See also* **boiling**, **evaporation**.

liquefied petroleum gas (LPG)   A mixture of **hydrocarbon** compounds which is used as a portable supply of gas fuel. It is made up mainly of **propane** (C_3H_8) and **butane** (C_4H_{10}) and these compounds are stored in metallic bottles under pressure.

lithium (Li)   A group I metal. It is the least reactive **element** in the group. Nevertheless it is stored under oil because of its reactivity towards air and water. It is soft, and can be cut with a knife, revealing a silvery surface which tarnishes readily. It has a steady reaction with water but reacts vigorously with acids.

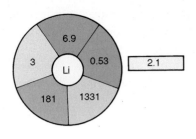

lithium

$$2Li(s) + 2H_2O(l) \rightarrow 2LiOH(aq) + H_2(g)$$

It is possible that the metal will become very important in the future if an economic process is developed by which to generate electricity by **nuclear fusion** reactions. The metal ion gives a characteristic red colour in a **flame test**.

litmus   A substance extracted from lichen that is used as an acid-base **indicator**. Litmus paper turns blue in the presence of an alkali and red in the presence of an acid.

acidic solution	pH 7	alkaline solution
red	purple	blue

litre   The unit of liquid volume in which 1 litre is equal to $1000\,cm^3$. The litre is commonly used in everyday life. However, in scientific uses, the cubic decimetre is used instead ($1000\,cm^3 = 1\,dm^3$).

lone pair of electrons   Pairs of outer shell electrons that are not used in chemical **bonds** within the compound. They can, however, be used for forming bonds with other compounds, as the example shows. *See also* **electronic configuration**.

lone pair of electrons *A lone pair bonding with a hydrogen ion (H+) to form an ammonium ion.*

Loudon forces   *See* **van der Waals' forces**.

LPG   *See* **liquefied petroleum gas**.

M

M *See* **molarity.**

M_r The **symbol** for the **relative molecular mass** and **relative formula mass** of a compound or molecule.

macromolecule A large molecule. The term is usually used for molecules with a **relative molecular mass** value greater than 1000, namely **polymers** such as **carbohydrates** and **plastics**, and molecules such as **proteins** and nucleic acids.

magnesium (Mg) A shiny grey group II metal. It is quite reactive, giving vigorous reactions with acids. It burns vigorously in air with a bright white light, hence its use in flares and fireworks. It also burns in carbon dioxide gas, and will react with steam to release hydrogen. The metal is obtained by the **electrolysis** of molten magnesium chloride. Much of this is extracted from seawater by the **Dow process.**

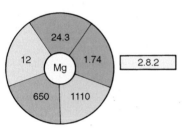

magnesium

The chief use of magnesium is in the production of low density **alloys** for use in the aircraft industry. Magnesium is also used in the production of steel, as a **reducing agent** in the chemical industry, and as **sacrificial anodes** in the protection of other metals.

magnesium compounds

Magnesium hydroxide $Mg(OH)_2$	Obtained by treating seawater with calcium hydroxide. It is used in pharmaceutical products such as *Milk of Magnesia* as a treatment for excess acidity in the stomach.
Magnesium carbonate $MgCO_3$	Found naturally as magnesite and as dolomite $(MgCO_3.CaCO_3)$ and is used for making heat-resistant (refractory) materials.
Magnesium chloride $MgCl_2$	After extraction from seawater the molten compound is electrolysed to produce magnesium metal.
Magnesium sulphate $MgSO_4.7H_2O$	*Epsom Salts.* These crystals are used to treat constipation. They are also used as a fire-proofing agent.

malachite A hydrated mineral form of copper(II) carbonate. Its formula can be written as either $Cu_2(OH)_2CO_3$ or $CuCO_3.Cu(OH)_2$.

malleable Able to be beaten or hammered into different shapes. The larger the **crystals** in a substance, the more malleable it is. Materials that are malleable are usually also **ductile**. Metals and alloys have these properties. *See also* **annealing**.

maltose ($C_{12}H_{22}O_{11}$) A **disaccharide** molecule. It is found as a breakdown product of **starch**. Maltose is a **glucose dimer**.

manganese (Mn) A **transition metal**. It is found naturally as the oxide (MnO_2) and is extracted either by **electrolysis** of the sulphate or by a **thermit reaction**. The chief use of the metal is in **alloys**, such as steels and bronzes.

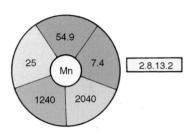

manganese

manganese compounds Manganese forms compounds where the metal has a valency of 2, 3, 4, 6 or 7.

Manganese(IV) oxide MnO_2	Important use in the dry battery. It is used as a catalyst and oxidizing agent.
Potassium manganate(VII) $KMnO_4$	A vivid purple crystalline substance which is a vigorous oxidizing agent. It is used as an antiseptic. It gives the MnO_4^- ion.

marble A **metamorphic rock**, a crystalline mineral form of calcium carbonate ($CaCO_3$). It is widely used as a decorative wall covering in buildings, and as a flooring material.

mass The amount of material a substance possesses. It is measured in grams (g) or kilograms (kg).

mass number (A) The sum of the number of **protons** and **neutrons** in the **nucleus** of the atom. This number is shown at the top left-hand side of the symbol when describing the **isotope**. The number below it is the **atomic number**. For example:

$$^1_1H \quad ^{12}_6C \quad ^{23}_{11}Na \quad ^{40}_{20}Ca \quad ^{208}_{82}Pb \quad ^{238}_{92}U$$

matches Small, thin sticks of wood with one end coated in chemicals that produce a **flame** when struck against a rough surface. The reaction is between phosphorus sulphide and potassium chlorate ($KClO_4$). These two chemicals are normally combined together in the match head, and when pulled across a rough surface (as on the match box), the **activation energy** is provided for the chemical reaction. This generates a lot of heat and a flame is produced.

In safety matches only one chemical is contained in the match head; the other (the phosphorus sulphide) is on the side of the box. This way, the matches cannot be lit accidentally.

melamine   A colourless, crystalline compound produced from **urea**. It undergoes **condensation polymerization** with **methanal** to produce melamine resins. These are heat- and light-resistant, colourless and shatter-resistant. They are widely used for picnicware and for kitchen worktops.

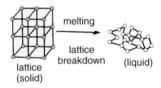

melamine

melting   The change of state from solid to liquid. This occurs when the energy of the particles is sufficient to break up the bonds holding them in a **lattice**. For a **pure** substance, melting occurs at a fixed temperature – the melting point. Some bonds remain but clusters of particles are mobile.

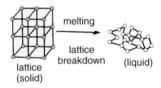

melting *Particles of a solid breaking up to form a liquid.*

melting point   The temperature at which a solid melts to form a liquid. It is therefore the temperature at which solid and liquid forms of the same substance (such as ice and water) are in **equilibrium**. At constant pressure, the melting point is a constant for a **pure** substance but it is lowered if impurities are added, hence ice can melt when sprinkled with **salt**. *See also* **freezing point**.

mercury (Hg) The only metal that is liquid at room temperature. It is used to make **amalgams**, in **electrolysis** cells as a **cathode**, and in thermometers. Its compounds and vapour are very poisonous.

mercury

metal Any chemical element such as iron or copper that is reactive, has a crystalline structure, is shiny and has high melting and boiling points.

Metals produce **cations**, react with acids, are strong and hard, are **malleable** and **ductile**, are good conductors of heat and electricity and react with non-metals.

Pure metals have few practical uses; most metals are used in the form of **alloys**.

metallic bond The chemical **bond** that exists between the atoms within metals. Metals contain positive **ions** and delocalized **electrons** which can move through the structure (a positively charged **lattice** and a cloud of electrons).

metalloid or semimetal An element that is neither a **metal** nor a **non-metal**. The best examples are silicon, germanium and arsenic. Such substances tend to have the physical properties of metals, in that they have a shiny appearance and high melting and boiling points but the chemical properties of non-metals, for example they do not react with acids. They are useful as **semiconductors**.

metamorphic rocks Rocks that have been formed from **igneous** or **sedimentary** rocks by high temperatures and/or pressures. Examples are **marble** and slate. *See also* **rock cycle**.

methanal A toxic gas belonging to the family of organic compounds called aldehydes. Methanal is produced by the **oxidation** of **methanol**. It is an important chemical in the production of **thermosetting polymers**. *See also* **melamine**, **phenol/methanal resins**, **urea/methanal resin**.

H_2CO

methanal

methane   A gaseous **alkane**. It is the main constituent of **natural gas**. It is also released from **petroleum** whilst it is being processed. It burns readily to give carbon dioxide and water and is an industrial source of hydrogen.

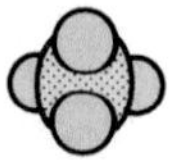

CH_4

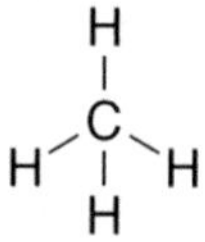

methane

methanol   An **alcohol**. It is a colourless, poisonous liquid. Methanol is produced in large quantities from **synthesis gas** at high temperature and pressure:

$$CO(g) + 2H_2(g) \rightarrow CH_3OH(g)$$

$$CO_2(g) + 3H_2(g) \rightarrow CH_3OH(g) + H_2O(g)$$

Methanol is used in several ways:
(a) as a solvent;
(b) to produce **ethanoic acid**;
(c) to produce **methanal** and **thermosetting polymers**;
(d) as an **additive** to **unleaded petrol**;
(e) in the production of **methylated spirit**.

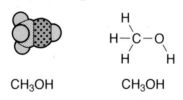

CH_3OH CH_3OH

methanol

methylated spirits   A mixture of **ethanol** (90%), **methanol** (9.5%), the **aromatic** hydrocarbon pyridine (0.5%) and traces of a purple dye. This mixing is done to make the ethanol unfit for drinking. There is no tax on the mixture and so it is inexpensive. It is used as a **fuel** and a **solvent**.

methylbenzene (formerly toluene)   An **aromatic** hydrocarbon that is produced from **petroleum**. It is used to make the explosive TNT (trinitrotoluene) and is a useful **solvent**.

methyl orange An **acid–base indicator**.

acid solution	alkaline solution
←	→
orange/red	yellow

methyl tertiary butyl ether (MTBE) *See* **unleaded petrol**.

mineral **1.** Any of the **inorganic** chemicals needed in the human diet. Examples are iron, calcium and potassium.
2. A naturally occurring crystalline form of an element or compound, which is used as a **raw material** in the chemical industry. The major minerals used in the UK are shown in the table. *See also* **ore**.

Mineral	Use
Clay	Building materials
Fluorspar	Plastics
Gypsum	Cement
Limestone	Cement, glass, paper, rubber, building materials
Phosphates	Fertilizers, detergents
Potash	Potassium chemicals, fertilizers
Rock salt	Detergents, soaps, food, textiles, paper, plastics, bleaches
Sand	Building materials, cement, glass, catalysts
Sulphur	Explosives, paints, plastics, dyes, fertilizers

mineral extraction The taking of minerals and **ores** from the ground. The main methods are:
(a) open-cast mining, for limestone, clay, coal, phosphate and iron ore.
(b) underground mining and solution mining for coal, tin, gold, uranium, rock salt and sulphur (*see also* **Frasch process**).

mineral processing The techniques of changing **minerals** and metallic **ores** into more useful forms. The standard steps in mineral processing are as follows:
(a) extraction
(b) crushing and grinding to a powder
(c) sorting according to size and properties
(d) removal of impurities
(e) chemical purification.

miscible   Capable of being mixed. Liquids that can mix together completely are described as miscible. They will **dissolve** in one other, for example water and ethanol, and will require **fractional distillation** to separate them. *See also* **immiscible**.

mixture   Two or more substances present in the same container. Examples are baking powder, air, gun-powder, petrol, methylated spirits. Mixtures have several properties:
(a) Mixtures can be made in all proportions.
(b) Making mixtures does not involve the release or absorption of heat.
(c) The properties of a mixture are the properties of all the components.
(d) Mixtures can be separated by physical means.

ml   Abbreviation for millilitre. 1000 ml = 1 **litre**.

mmHg   A unit of pressure equal to that exerted by a height of one millimetre of mercury (Hg). One **atmosphere** pressure equals 760 mmHg. The unit has been largely replaced by the pascal (1 mmHg = 133.3 pascals).

Mohs scale   A scale used to express the hardness of solids by comparing them against ten standards. The hardness of **minerals** varies widely and so a scale such as this can be useful in helping to identify unknown specimens. On the scale, hardness increases from 1 to 10.

Hardness	Standard mineral
1	Talc
2	Rock salt or gypsum
3	Calcite
4	Fluorite
5	Apatite (or window glass)
6	Feldespar
7	Quartz or flint
8	Topaz
9	Corundum
10	Diamond

Mohs scale

molarity (M)   The **concentration** of a **solution** expressed in **moles** per cubic decimetre of solution (mol dm^{-3}). A solution containing 2 moles per dm^3 is expressed as 2 M. *See also* **molar solution**.

molar solution   A **solution** in which the **concentration** of **solute** is one **mole** per cubic decimetre of solution (1 mol dm^{-3}). In other words, a solution where one mole of solute has been dissolved in the **solvent** and then sufficient solvent has been added to make 1000 cm^3 of solution.

mol dm^{-3}   A unit of **concentration**. 1 mol dm^{-3} is equal to one **mole** of substance in one cubic decimetre (dm^3) of solution. The unit is sometimes written mol/dm^3 and is often given the symbol M, for example 2 M. *See also* **molarity**.

mole   The **SI unit** of amount of substance, equal to the amount of a substance (element or compound) that contains L particles, where L is the **Avogadro constant**.
   L is defined as the number of atoms there are in 12 g of the carbon-12 **isotope**, and is equal to 6.022×10^{23}. It follows that the mass of one mole of an element (or compound) is the **relative atomic mass**, A_r (or **relative molecular mass**, M_r), which is expressed in grams. For example, one mole of

oxygen	(O_2)	is 16 + 16	= 32g
water	(H_2O)	is 2(1) + 16	= 18g
ethene	(C_2H_4)	is 2(12)	= 28g.

To calculate the number of moles in an amount of substance, divide its **mass** (in grams) by its A_r or M_r value. *See also* **concentration**.

molecule   The smallest particle of an **element** or **compound** that exists independently and has the properties characteristic of that substance. A molecule contains atoms bonded together in a fixed whole number ratio, for example:

Oxygen O_2	Nitrogen N_2
Water H_2O	Carbon dioxide CO_2
Sulphur S_8	Hydrogen chloride HCl

All organic compounds except **polymers** are molecules.

molecular formula   A formula that shows the number and types of atoms in a molecule. It tells us nothing about how the atoms are arranged. For example, $C_4H_{10}O$ is the molecular formula of both butanol and ether (ethoxyethane). *See also* **empirical formula**, **structural formula**.

monatomic molecule   A **molecule** that contains only one **atom**. Only the **noble gases** are monatomic. Because of their **electronic configurations** it is difficult for noble gas atoms to form chemical bonds, and so they exist in the monatomic form.

monoclinic sulphur   *See* **sulphur**.

monobasic acid   An **acid** that contains only one hydrogen atom per molecule which can be replaced by a metal. Only normal **salts** can be formed by monobasic acids – not **acid salts**. Examples include hydrochloric acid (HCl) and nitric acid (HNO_3).

monomer   Any compound from which **polymers** are made. For example **ethene** $CH_2=CH_2$ forms **poly(ethene)** $(CH_2–CH_2)_n$.

monosaccharide   The simplest kind of **sugar**. These compounds contain a single molecule, as opposed to two (or more) molecules that have reacted together to form **disaccharides** or **polysaccharides**. Examples are glucose ($C_6H_{12}O_6$), fructose ($C_6H_{12}O_6$), and ribose ($C_5H_{10}O_5$).

mortar   A mixture of water, sand and cement that is used to bond building materials.

multiple bonds   Chemical **bonds** that contain four or six electrons. **Double bonds** contain four electrons (two shared pairs) and **triple bonds** contain six (three shared pairs). Examples of compounds that contain multiple bonds are:

$$\text{ethene } C_2H_4 – \text{double bond}$$

$$\text{ethyne } C_2H_2 – \text{triple bond}$$

See also **single bonds**.

N

nanometre (nm) A unit of length. It is 10^{-9} of a metre, that is 1,000,000,000 nm make 1 m. The nanometre is useful in measuring the length of chemical **bonds**, for example the H–H bond length is equal to 0.074 nm.

naphtha A mixture of **hydrocarbons** that is produced by the **fractional distillation** of **petroleum**. The boiling points of the compounds in the mixture are in the range 80–160 °C. Naphtha is usually subjected to **cracking**. **Ethene** is an important product of this process. Naphtha can be oxidized to give **ethanoic acid**.

native (used of an element) Found in nature uncombined with any other element.

natural gas A mixture of mainly **hydrocarbon** gases which is found in deposits beneath the Earth's surface. **Natural gas** and **petroleum** are often found together. **Methane** is generally the major constituent. The inert gas, helium, is sometimes present in the mixture. Natural gas is used as a **fuel** in both industry and the home.

negative electrode *See* **cathode**.

neon A **noble gas** that is present in very small amounts in the Earth's atmosphere. It is an inert gas and forms no known compounds. It is used to fill fluorescent tubes.; a red glow is produced when an electric current is passed through the gas. Neon lights are often used for advertising signs.

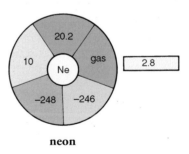

neon

neutral particle One that does not have an electric charge, such as a **neutron**.

neutral oxide An **oxide** that does not react with either **acids** or **alkalis**. The best example is water.

neutral solution A **solution** that is neither acidic (*see* **acid**) nor **alkaline**. Neutral solutions contain the same concentration of **hydroxide ions** and **oxonium ions**. The **pH** of a neutral solution is 7 at 25 °C.

neutralization The process in which either the **pH** of an acidic solution (*see* **acid**) is increased to 7 or the pH of an **alkaline** solution is decreased to 7. The resulting solution contains the same concentration of **oxonium ions** and **hydroxide ions**; in other words, it is neutral.

An acid can be neutralized by the addition of a base or a compound such as a carbonate.

$$HCl(aq) + NaOH(aq) \rightarrow NaCl(aq) + H_2O(l)$$

$$2HCl(aq) + CaCO_3(s) \rightarrow CaCl_2(aq) + CO_2(g) + H_2O(l)$$

Acid–base **indicators** show when neutralization is complete.

neutron One of the particles found in the **nucleus** of all **atoms** except hydrogen. It has approximately the same mass as the **proton** but no charge. *See also* **nuclear reactions**.

nickel (Ni) A **transition metal**. It is a magnetic substance that occurs as the **sulphide** and is oxidized to the **oxide**, reduced by hydrogen and then purified using carbon monoxide gas:

$$Ni(s) + 4CO(g) \rightarrow Ni(CO)_4(g)$$

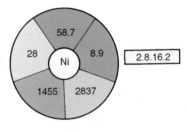

This reaction can be reversed at high temperatures, producing pure nickel. Nickel can have a **valency** of 2, 3 or 4, but only nickel(II) salts are common. Nickel is used as a **catalyst**, in **alloys** (Nichrome, coinage metal, stainless steel), in plating and in **cells**.

nickel

nitrate A **salt** of **nitric acid**, containing the $-NO_3$ group and with a **valency** of 1. A nitrate can be made by treating a metal **carbonate** or **oxide** with the dilute acid:

$$CuCO_3(s) + 2HNO_3(aq) \rightarrow Cu(NO_3)_2(aq) + CO_2(g) + H_2O(l)$$

$$ZnO(s) + 2HNO_3(aq) \rightarrow Zn(NO_3)_2(aq) + H_2O(l)$$

Nitrates are easily decomposed by heat. There are three kinds of reaction:
(a) $2KNO_3(s) \rightarrow 2KNO_2(s) + O_2(g)$ examples: sodium, potassium.
(b) $2Cu(NO_3)_2(s) \rightarrow 2CuO(s) + O_2(g) + 4NO_2(g)$ all metal nitrates other than shown in (a) and (c).
(c) $2AgNO_3(s) \rightarrow 2Ag(s) + 2NO_2(g) + O_2(g)$ examples: silver, mercury.
Nitrates are important **fertilizers**, particularly sodium nitrate and ammonium nitrate. If used incorrectly they can lead to serious water **pollution**.

nitric acid (HNO_3)   A colourless, corrosive liquid made from ammonia by **oxidation** over a platinum/rhodium **catalyst**. There are three stages:

$$4NH_3(g) + 5O_2(g) \rightarrow 4NO(g) + 6H_2O(g)$$

$$4NO(g) + 2O_2(g) \rightarrow 4NO_2(g)$$

$$4NO_2(g) + 2H_2O(l) + O_2(g) \rightarrow 4HNO_3(aq)$$

The acid is distilled to a concentration of 68% and can be prepared in the laboratory by heating a **nitrate** with concentrated sulphuric acid:

$$KNO_3(s) + H_2SO_4(l) \rightarrow KHSO_4(s) + HNO_3(g)$$

Nitric acid is a vigorous **oxidizing agent** used in the production of **fertilizers, explosives, dyes** and the polymer **nylon**. The **salts** of nitric acid are termed nitrates. *See also* **nitrogen oxides, nitrites**.

nitrites   **Salts** of nitrous acid (HNO_2). The sodium and potassium salts can be made by heating the nitrate.

$$2NaNO_3(s) \rightarrow 2NaNO_2(s) + O_2(g)$$

They are used in the curing of meats.

nitrogen (N)   A non-metallic element in group V of the **periodic table**. It is an unreactive, diatomic gas (*see* **diatomic molecule**) that forms about 78% of the Earth's atmosphere. It is produced by the **fractional distillation** of liquid air.

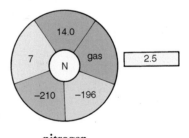

nitrogen

Nitrogen is an important chemical because of the need for **nitric acid** and **ammonia**. In the **Haber process**, nitrogen is fixed from the air to make ammonia. *See also* **fertilizer**.

Nitrogen gas is used to provide an **inert** atmosphere, for example, in **annealing** steel; silicon chip production; food packaging; and glass production.

Liquid nitrogen is used in the refrigeration of such items as medical samples and in the transportation of food. *See also* **nitrogen oxides, nitrogen cycle**.

nitrogen cycle   The process that shows the ways in which animals, plants and humans are involved in the chemical links between **nitrogen, ammonia** and **nitrates**. Nitrogen is fixed from the atmosphere by

- action by soil bacteria;
- electrical discharges (lightning);
- the **Haber process.**

The nitrates in the soil are taken up by plants, which are then eaten by animals. Both animal and plant **protein** eventually rot and revert to ammonia. Animals also excrete urine, which decomposes to ammonia.

Nitrogen is returned to the atmosphere by the action of other bacteria.

Despite human extraction of nitrogen from the air by the Haber process the amount of nitrogen in the atmosphere remains approximately constant.

See also **acidification, greenhouse gases.**

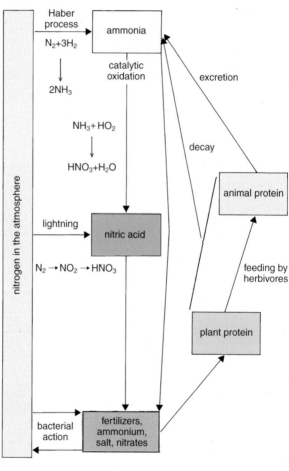

nitrogen cycle

nitrogen oxides

Nitrogen monoxide NO	A colourless gas that can be produced by the produced by the action of moderately concentrated nitric acid on copper. The gas reacts immediately with oxygen to form nitrogen dioxide: $2NO(g) + O_2(g) \rightarrow 2NO_2(g)$.
Nitrogen dioxide NO_2 (nitrogen(IV) oxide)	A brown gas. It has a choking smell and irritates the lungs and windpipe. Breathing it can lead to death from pneumonia. It can be produced in the laboratory by the action of concentrated nitric acid on copper.
Nitrous oxide N_2O (nitrogen(I) oxide)	A colourless gas with a sweetish smell. It is an anaesthetic and is used in dentistry. It is sometimes called laughing gas.

nm   *See* **nanometre**, *Appendix A*.

noble gases **or** inert gases   Elements in group 0 of the **periodic table**. They are all very unreactive monatomic gases (*see* **monatomic molecule**) that are present in small amounts in the atmosphere.

Their **electronic configurations** lead to very stable chemistry and the elements show no tendency to lose or gain electrons. Because of this, they find it hard to form compounds. Compounds do exist of the more massive elements, for example XeF_4, but these are rare.

All the gases are produced commercially by the **fractional distillation** of liquid air, except helium which is recovered from **natural gas**. *See also* **neon**, **argon**, **krypton**, **xenon**.

noble gas structure   The atomic structure of **noble gases** in which there are eight electrons in the outer shell of the atom. This is a stable **electronic configuration** because the eight electrons completely fill a part of the shell making it difficult to add another or take one away. When elements form **ions**, the ions that are formed have the noble gas configuration, as shown below.

Element	Ion	Electronic configuration atom	ion	Noble gas equivalent
Sodium	Na^+	2·8·1	28	Neon
Fluorine	F^-	2·7	2·8	Neon
Calcium	Ca^{2+}	2·8·8·2	2·8·8	Argon
Sulphur	S^{2-}	2·8·6	2·8·8	Argon

non-conductor   *See* **insulator**.

non-metal   An **element** that either (i) has a molecular structure, with low melting and boiling points and is thus a gas at room temperature (for example, oxygen, nitrogen, bromine) or (ii) is a **giant structure** with **covalent bonding** and therefore has high melting and boiling points; examples are carbon, silicon. Typical properties of non-metals are as follows:
(a) they have poor **conductivity**;
(b) they do not react with acids;
(c) they produce **acidic oxides**;
(d) they form **covalent compounds**;
(e) they give rise to **anions**.
Examples include the **halogens**, the **noble gases**, oxygen, sulphur, carbon and nitrogen.

NPK   *See* **fertilizer**.

nuclear reaction   A **reaction** in which changes occur in the **nucleus** of an atom and a new element can be formed. Most elements have both stable and unstable (radioactive) **isotopes**; for example carbon has the stable isotopes carbon-12 and carbon-13 and the unstable one carbon-14. However some elements, such as uranium and plutonium, have no stable isotopes. Whether or not an isotope is unstable depends on the numbers of **neutrons** and **protons** in the nucleus. When an isotope is unstable, several reactions can occur:
(a) The isotope can split – **fission** can occur. This creates two stable isotopes. Neutrons are released and so is a great deal of energy.
(b) A neutron can decay into a proton and an electron:

$$\frac{1}{0}n \rightarrow \frac{1}{1}p + \frac{0}{-1}e$$

The electron (**beta particle**) is expelled from the nucleus and the atomic number of the atom increases by 1, for example:

$$\frac{209}{82}Pb \rightarrow \frac{209}{83}Bi + \frac{0}{-1}e$$

This creates a new element. There is no change in mass.
(c) An **alpha particle** is expelled from the atom:

$$\frac{238}{92}U \rightarrow \frac{234}{90}Th + \frac{4}{2}He$$

(d) An unstable atom can emit **gamma rays**. This often occurs after the emission of a beta particle. *See also* **half-life**.

These nuclear reactions may be used in many ways. Nuclear power stations use the heat produced in the fission of uranium, plutonium or thorium isotopes. The heat turns water into steam which drives a turbine producing electricity. This same source of energy is also used in atomic weapons. These have terrible destructive powers.

Isotopes producing **gamma rays** are used to destroy bacteria in food processing and cancerous cells in the body. Isotopes are also used in industry for a variety of analytical purposes, such as detecting cracks in pipelines.

The products of radioactive decay are dangerous as they can produce harmful effects on the body including leukaemia and cancers. Great care has to be taken when using radioactive materials.

nucleons Particles found in the **nucleus** of the **atom**, that is, **protons** and **neutrons**.

nucleus The part of an **atom** where the mass is concentrated. It contains **protons** and **neutrons** and is usually pictured as being the compact centre of a spherical atom. **Electrons** move around the nucleus. The nucleus has a positive charge and in the neutral atom this is balanced by the charges on the electrons.

The hydrogen-1 **isotope** is the only atom whose nuclei contain no neutrons. *See also* **nuclear reactions**.

a nucleus of carbon-12
contains 6 protons
and 6 neutrons

a nucleus of carbon-14
contains 6 protons
and 8 neutrons

nucleus *Different isotopes showing different nuclei.*

nuclide An atomic **nucleus**.

nylon A family of synthetic **polyamide** polymers. The most common of them is nylon 6.6, a **fibre** with great strength (see figure overleaf). Its uses include fabrics (shirts, cloth), yarns (tights, knitwear), carpets, ropes and nets. Nylon is useful because it will not rot, it does not absorb water but it does stretch. This is useful in tights, and in ropes used by climbers. It is often mixed with other fibres, such as wool, to get the correct balance of properties.

nylon *The most common nylon, nylon 6.6.*

O

octane   A **hydrocarbon** with the formula C_8H_{18}. *See also* **alkanes, octane rating**.

octane rating **or** octane number   A measure of the anti-knock qualities of **petrol**. Mixtures of petrol and air have to explode at exactly the correct moment in an internal combustion engine. If the wrong kind of fuel is used, ignition of the mixture can occur before it should. This leads to the characteristic sound of *knocking* (or pinking).
   The higher the octane rating, the better the fuel. Unleaded petrol has an octane rating of 98. The greater the proportion of the **hydrocarbon** molecules in the fuel that have branched hydrocarbon chains, the higher the rating will be.
   Octane ratings can be raised by adding substances such as tetraethyl lead. This caused lead pollution in the atmosphere and is now no longer used. *See also* **unleaded petrol**.

oil   Any of a group of **hydrocarbons** that may be solid (**fats** or waxes) or liquid. There are three types of oils. *Fixed oils* are found in animals and plants, and are used in foods, soaps and paint. *Mineral oils* are obtained from refining petroleum and are used as fuels and lubricants. *Essential oils* are derived from certain plants which possess pleasant odours, and are used in perfumes and flavourings.

olefin   *See* **alkenes**.

oleum   A solution of sulphur(VI) oxide in concentrated sulphuric acid. It is a very corrosive substance and is a vigorous **oxidizing agent**. *See also* **contact process**.

orbital   *See* **electronic configuration**.

ore   A naturally occurring substance from which an **element** can be extracted. Examples are shown overleaf. *See also* **mineral**.

Element	Ore
Aluminium	Bauxite
Chlorine	Rock salt
Copper	Chalcopyrite ($CuFeS_2$)
Iron	Haematite (Fe_2O_3)
Lead	Galena (PbS)
Zinc	Zinc blende (ZnS)

ore *Elements and the ores from which they are extracted.*

organic Originally a term meaning produced by, or found in, plants or animals. The term has now been extended to refer to all compounds containing both carbon and hydrogen, except the **hydrogencarbonates**.

organic chemistry The study of the compounds of carbon. It does not include carbonates, carbon dioxide, etc. *See also* **inorganic chemistry**.

oxidation A chemical reaction in which one of the following occurs to a substance:
(a) it gains oxygen: $2Mg(s) + O_2(g) \rightarrow 2MgO(s)$
(b) it loses hydrogen: $CH_4(g) + Cl_2(g) \rightarrow CH_3Cl(l) + HCl(g)$
(c) it loses electrons: $Cu(s) \rightarrow Cu^{2+}(aq) + 2e^-$
See also **oxidizing agent**, **reduction**, **redox**.

oxide A compound formed between an element and oxygen only. Oxides can be formed by:
(a) direct combustion of the elements;
(b) oxidizing compounds (*see* **oxidation**);
(c) the action of heat on a **carbonate**, a **hydroxide** or certain **nitrates**.
Oxides can be reduced to the element with a suitable **reducing agent**, e.g. hydrogen, carbon or carbon monoxide. *See also* **acidic oxide**, **amphoteric**, **basic oxide**.

oxidizing agent A substance that causes the **oxidation** of another substance. When this happens the oxidizing agent is itself reduced (*see* **reduction**).
   Common oxidizing agents are oxygen O_2, chlorine Cl_2, ozone O_3 and hydrogen peroxide H_2O_2.
   Examples of oxidizing agents at work include:

$$2Mg(s) + O_2(g) \rightarrow 2MgO(s)$$

$$H_2O_2(aq) + H_2S(g) \rightarrow 2H_2O(l) + S(s)$$

oxonium ion (H₃O⁺) An **ion** formed when a positive hydrogen ion (i.e. a **proton**) bonds to a water molecule. More than one water molecule is probably involved but, for simplicity only one is usually shown:

$$H^+ \cdots O \overset{H}{\underset{H}{\diagup\diagdown}} \quad H_3O^+(aq)$$

The oxonium ion can be used in ionic equations in place of $H^+(aq)$:

$$H_3O^+(aq) + OH^-(aq) \rightarrow 2H_2O(l)$$
$$\text{acid} \qquad\quad \text{alkali}$$

oxygen (O) A gaseous, non-metallic element in group VI of the **periodic table**. It makes up 21% of the Earth's atmosphere. It is a vigorous **oxidizing agent** and is vital for the **respiration** of plants and animals.

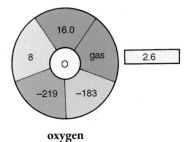

oxygen

Oxygen is a reactive gas, readily forming **oxides** with most elements. It is obtained industrially by the **fractional distillation** of liquefied air and in the laboratory by the decomposition of hydrogen peroxide:

$$2H_2O_2(aq) \rightarrow 2H_2O(l) + O_2(g)$$

A **catalyst** such as manganese(IV) oxide is usually used.

Oxygen is a colourless, odourless diatomic gas (*see* **diatomic molecule**). It is neutral and is only slightly soluble in water (40 cm³ per cubic decimetre). However, it is sufficiently soluble to allow fish and other aquatic life to live in water. Problems arise if the oxygen in water drops to too low a level. Then, fish suffocate and die. Such a drop in the concentration of oxygen is usually caused by pollution. *See also* **eutrophication**.

The chemical test for oxygen is to plunge a glowing splint into the gas. If the gas is oxygen, the splint will relight. The gas will also make anything that is already burning in air burn much more fiercely.

Oxygen is used extensively in steel-making and welding, as a rocket propellant together with kerosene or hydrogen, and for life-support systems in medicine or breathing apparatus. It is also used to produce other chemicals.

ozone (O₃)   An **allotrope** of oxygen where the molecule is triatomic $O_3(g)$. It is a blue gas and is a very powerful **oxidizing agent** and extremely poisonous. It can be formed from oxygen by passing a spark through the gas:

$$3O_2(g) \rightarrow 2O_3(g)$$

Ozone is a powerful **germicide**, and its great oxidizing power makes it useful, in very high dilution, for ventilating spaces where fresh air has limited access, for example in underground railways. It is also used in bleaching textiles, paper and oil, and in purifying the water in swimming pools. Ozone is found in the stratosphere. *See also* **Earth's atmosphere, ozone layer**.

ozone layer   Part of the **Earth's atmosphere** that contains **ozone**. About one molecule in every 10,000,000 in the Earth's atmosphere is ozone. Although this doesn't sound a lot, because the atmosphere is so large there are billions of tonnes of the gas, which is concentrated in a region 20 km thick round the Earth in the stratosphere. Here it provides protection against **ultraviolet (UV) radiation** from the Sun. The ozone molecules stop most of the UV reaching the Earth where it would cause great damage to crops and marine life and to human skin.

The UV breaks down the ozone into oxygen, but ozone is continuously reformed by the action of UV radiation on oxygen. Every day, on average, 300 million tonnes of ozone are destroyed and produced in the stratosphere.

Since 1980, the amount of ozone in the stratosphere has been reduced, particularly over the poles where the coverage can become thin during the spring. This reduction is blamed on atmospheric pollutants such as **CFCs**. *See also* **greenhouse effect, greenhouse gases, pollution**.

P

paint A chemical mixture that is used to protect and decorate materials. All paints contain:
(a) **pigments** to provide colour;
(b) binders to dissolve the other materials;
(c) thinners to make the paint easy to apply.
Other **additives** might include:
(d) drying agents to speed up the drying process;
(e) silicones to give resistance against the weather;
(f) **biocides** to protect against fungal attack;
(g) extenders to provide bulk to the paint.

paraffin A **fuel** obtained from **petroleum**. It has a boiling range of 160–250 °C.
   Paraffin wax consists of a mixture of solid **hydrocarbons**. It is used to make candles. *See also* **kerosine**.

particle The general term used to describe extremely small piece of matters. All materials are made up of particles, which can be **atoms, ions, molecules** or **polymers**.

particle theory *See* **kinetic theory**.

pascal The **SI unit** of pressure. One **atmosphere** pressure is about 100 kilopascals (100 kPa). One pascal is equivalent to a force of 1 newton on 1 square metre ($1\,N\,m^{-2}$).

passive (used of a metal) Possessing a surface layer of **oxide** which makes it unreactive. Aluminium has such a layer and iron can be given one by dipping the metal into concentrated nitric acid.

pasteurization A heat-treatment process to reduce the number of microorganisms present in food in order to extend its shelf life. Milk is pasteurized by being heated to 71.7 °C for at least 15 seconds. Pasteurized milk will keep in the fridge for four to five days. *See also* **UHT, sterilization**.

peat A **fossil fuel**.

peptide A compound formed when two or more **amino acids** react together. If three or more amino acids are involved the term **polypeptide** is

peptide *Peptide link*

usually used. **Proteins** consist of long polypeptide chains that may be linked together in a variety of ways. Peptides contain the *peptide link*.

percentage composition The proportion of the different **elements** making up a **compound**. This is usually expressed in percentage terms.

Knowing the percentage composition of a compound it is possible to work out its **empirical formula**. Say, for example, a compound is known to have the following percentage composition: Ca 40%, C 12%, O 48%. To find the *molar* ratio of the elements in the compound, divide each figure by the respective **relative atomic mass** for that element: (Ca) $\frac{40}{40} = 1$ (C) $\frac{12}{12} = 1$ (O) $\frac{48}{16} = 3$. The molar ratio is 1:1:3, therefore the ratio of the atoms is 1:1:3 and the empirical formula is $CaCO_3$.

	% by mass of:	
Element	carbon	hydrogen
Methane CH_4	75	25
Ethyne C_2H_2	92	8

percentage composition

period A horizontal row of **elements** in the **periodic table**, containing a sequence of elements of consecutive **atomic number** and steadily changing chemical properties. There are seven periods in all. *See also* **group**.

periodic table A way of presenting all the **elements** so as to show their similarities and differences.

The elements are arranged in increasing order of **atomic number (Z)** as you go from left to right across the table. For example:

Period 3

Na	Mg	Al	Si	P	S	Cl	Ar
11	12	13	14	15	16	17	18

The horizontal rows are called **periods** and the vertical columns, **groups**. The group found at the right hand side of the periodic table consists of the **noble gases**:

Period 1	He
2	Ne
3	Ar
4	Kr
5	Xe
6	Rn

There is a progression from **metals** to **non-metals** across each period.

periodic table

groups	1	2						3	4	5	6	7	0(8)
periods													
1													2 **He** helium 4.0
2	3 **Li** lithium 6.9	4 **Be** beryllium 9.0						5 **B** boron 10.8	6 **C** carbon 12.0	7 **N** nitrogen 14.0	8 **O** oxygen 16.0	9 **F** fluorine 19.0	10 **Ne** neon 20.2
3	11 **Na** sodium 23.0	12 **Mg** magnesium 24.3						13 **Al** aluminium 26.9	14 **Si** silicon 28.1	15 **P** phosphorus 31.0	16 **S** sulphur 32.1	17 **Cl** chlorine 35.5	18 **Ar** argon 39.9

transition metals

	3	4	5	6	7						3	4	5	6	7	0(8)
4	21 **Sc** scandium 45.0	22 **Ti** titanium 47.8	23 **V** vanadium 50.9	24 **Cr** chromium 52.0	25 **Mn** manganese 54.9	26 **Fe** iron 55.9	27 **Co** cobalt 58.9	28 **Ni** nickel 58.7	29 **Cu** copper 63.5	30 **Zn** zinc 65.4	31 **Ga** gallium 69.7	32 **Ge** germanium 72.6	33 **As** arsenic 74.9	34 **Se** selenium 79.0	35 **Br** bromine 79.9	36 **Kr** krypton 83.8
5	39 **Y** yttrium 88.9	40 **Zr** zirconium 91.2	41 **Nb** niobium 92.9	42 **Mo** molybdenum 95.9	43 **Tc** technetium 98	44 **Ru** ruthenium 101.1	45 **Rh** rhodium 102.9	46 **Pd** palladium 106.4	47 **Ag** silver 107.9	48 **Cd** cadmium 112.4	49 **In** indium 114.8	50 **Sn** tin 118.7	51 **Sb** antimony 121.8	52 **Te** tellurium 127.6	53 **I** iodine 126.9	54 **Xe** xenon 131.3
6	57 **La** lanthanum 138.9	72 **Hf** hafnium 178.5	73 **Ta** tantalum 181.0	74 **W** tungsten 183.9	75 **Re** rhenium 186.2	76 **Os** osmium 190.2	77 **Ir** iridium 192.2	78 **Pt** platinum 195.1	79 **Au** gold 197.0	80 **Hg** mercury 200.6	81 **Tl** thallium 204.4	82 **Pb** lead 207.2	83 **Bi** bismuth 209.0	84 **Po** polonium 209	85 **At** astatine 210	86 **Rn** radon 222

Group 1:
- 19 **K** potassium 39.1
- 37 **Rb** rubidium 85.5
- 55 **Cs** caesium 132.9

Group 2:
- 20 **Ca** calcium 40.1
- 38 **Sr** strontium 87.6
- 56 **Ba** barium 137.3

1 **H** hydrogen 1.0

key:
atomic no. **Symbol**
name
relative atomic mass

metal	non metal	transition metal	metalloid

Similar elements are found within a group. These elements have a similar **electronic configuration**, e.g.:

| group I | alkali metals | Li | Na | K | |
| group VII | halogens | F | Cl | Br | I |

The number of **electrons** in the outer shell is the same as the number of the group, for example group I.

lithium	2·1
sodium	2·8·1
potassium	2·8·8·1

The elements in the block between groups II and III are called the **transition metals**. These have many similarities: they produce coloured compounds, have variable **valency** and are often used as **catalysts**. Elements 58 to 71 are known as *lanthanide* or rare earth elements. These elements are found on Earth in only very small amounts.

Elements 90 to 103 are known as the *actinide* elements. They include most of the better known radioactive elements (those that take part in **nuclear reactions**). The elements with atomic numbers greater than 92 do not occur naturally. They have all been produced artificially by bombarding other elements with particles. Plutonium (atomic number 94) is formed in nuclear reactors.

Elements with atomic numbers of 104 and above continue to be named after famous scientists or places. For example, element 104 is rutherfordium (Rf), and 109 is meitnerium (Mt).

permanently hard water *See* **hard water**.

peroxide A compound containing the O_2^{2-} ion or $-O_2$ group. The best example is hydrogen peroxide (H_2O_2). The **alkali metals** also form peroxides. Hydrogen peroxide can be made by the action of acid on these:

$$Na_2O_2(s) + 2HCl(aq) \rightarrow 2NaCl(aq) + H_2O_2(aq)$$

Peroxides are vigorous **oxidizing agents**. Hydrogen peroxide can be used in dilute solution as a **disinfectant** and a **bleach**.

Perspex Trade name for **poly(methylmethacrylate)**, a **polymer**.

pesticide *See* **biocide**.

petrochemical Any chemical that has been made from **petroleum**. Examples include **ethene** and **propene**. Petrochemicals are used in the manufacture of other chemicals, e.g. **poly(ethene)** and **poly(propene)**, i.e. they are *intermediates* in the production of finished products.

petrol or **gasoline** A mixture of **hydrocarbons** that is used as a fuel in internal combustion engines. It is produced from **petroleum** in a refinery. It is principally a mixture of C_5-C_{10} **alkanes** obtained by both straight **fractional distillation** and by **cracking** and **reforming**. *See also* **octane rating, unleaded petrol**.

petroleum The complex mixtures of **hydrocarbons** that is found in the Earth's crust as, for example, **natural gas** and crude **oil**. Petroleum is formed over millions of years from the remains of marine animals and plant organisms. It is the raw material for the petrochemical industry and is the source of our **petrol, diesel** fuel, heating oil, fuel oil and gas supplies.

Vast reserves of petroleum are found in the Middle East, the United States, Russia, Georgia, Central America and the North Sea. However, the supply of petroleum will not last forever and the search is now on for the substance that will replace it in our lives.

How petroleum is converted into useful products is described under **refining**. *See also* **fractional distillation**.

pH A scale for measuring the acidity or alkalinity of a **solution** (*see* **acid, alkali**).

The lower the value, the more acidic the solution; i.e. the greater the concentration of **oxonium ions** there is within it. A **neutral solution**, where the concentrations of oxonium and hydroxide ions are equal, has a pH of 7 at 25 °C.

acidity increasing	neutral	alkalinity increasing
←		→
0	3	7	10	14

The pH of a solution is the negative logarithm (base 10) of the concentration of oxonium ions (**mol dm^{-3}**). Thus,

$$pH = -\log_{10}[H_3O^+]$$

For example, the pH of a solution whose concentration is 0.1 mol dm^{-3} is $-\log(0.1) = 1$.

Substance at a concentration of 1 mol/dm^3	pH
Strong acid (HCl)	0
Weak acid (vinegar) (CH$_3$COOH)	3
Pure water	7
Ammonia solution	11
Strong alkali (NaOH)	14
Stomach fluid	1
Soda water	4.5
Cow's milk	6
Sea water	8.5

phase change *See* **change of state**.

phenol A colourless, crystalline, **aromatic** solid, which turns pink on exposure to air and light.

Phenol is made by reacting **benzene** with **propene** (CH$_3$CH=CH$_2$) and then oxidizing the product. It is a very corrosive, poisonous chemical. 35% of phenol production is used in the production of **phenol/methanal resins** and a further 27% in the production of the transparent **poly(carbonate)** materials and **epoxy resins.** It is also used in the manufacture of **nylon,** dyes and detergents. Disinfectants and many household products are based on molecules similar to phenol, i.e. compounds with similar (or better) disinfectant properties but which are less corrosive, e.g. TCP and Dettol.

The O–H group in phenol has an acidic hydrogen and **salts** can be formed, such as sodium phenolate, C$_6$H$_5$O$^-$Na$^+$(s).

C$_6$H$_5$O$^-$Na$^+$(s)

O–H

C$_6$H$_5$OH

phenol

phenol/methanal resins Dark coloured, **thermosetting, condensation polymers** made from **phenol** and **methanal**. The polymers have been used for electrical fittings because of their good electrical insulating properties, and for such items as saucepan handles because of their poor conduction of heat. The first **plastic** materials to be made were produced from these resins but, for everyday items, their use has largely been replaced by similar materials that can be coloured. They are still used for rocket nose cones and other heat shields. *See also* **melamine, urea/methanal resins.**

phenolphthalein An **indicator** used to follow **acid–base** reactions.

acid solution   pH7   alkaline solution

colourless   red

phenolphthalein

phosphates **Salts** of **phosphoric(V) acid**. Phosphoric acid is a tribasic acid (it has three replaceable hydrogen atoms) and, therefore, gives rise to three kinds of salts. For example, with sodium (Na) it can form the salts:
(a) sodium phosphate Na_3PO_4
(b) sodium hydrogenphosphate Na_2HPO_4
(c) sodium dihydrogenphosphate NaH_2PO_4
Phosphates are used in **fertilizers** to replace the phosphorus-containing compounds in the soil. They come in several forms.

Superphosphate is a mixture of calcium sulphate and calcium dihydrogenphosphate $Ca(H_2PO_4)_2$. A mixture of ammonium nitrate and ammonium hydrogenphosphate $(NH_4)_2HPO_4$ is made by reacting ammonia with a mixture of phosphoric(V) and nitric acids. Calcium phosphate is the chief constituent of animal bones and phosphates have been used extensively in washing powders and **detergents**, although this use is now avoided as it caused water pollution (*see* **eutrophication**). **Slag** from **blast furnaces** contains calcium phosphate and may be put directly onto the soil.

Phosphate ions produce a bright yellow solution when mixed with a solution of ammonium molybdenate/nitric acid. This is the test for phosphate ions.

phosphoric(V) acid (H_3PO_4) A tribasic acid (it has three replaceable hydrogen atoms) that gives rise to three kinds of **phosphate** salts. It is a solid at room temperature but is usually sold as a viscous solution in water. It is used for rust-proofing steel by forming a protective layer of iron phosphate. It is also used in the food and drug industries but most (90%) of the production goes to produce **fertilizers**.

phosphorus (P) A solid non-metallic element in group V of the **periodic table**. Three **allotropes** exist: white, red and black phosphorus. The information in the chart below refers to white phosphorus. The allotropes have very different physical properties, including crystal structure, density, melting and boiling points.

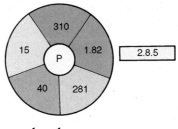

phosphorous

White phosphorus is very poisonous and has to be kept under water because it bursts into flame when it comes into contact with air, forming the **oxide**:

$$P_4(s) + 5O_2(g) \rightarrow P_4O_{10}(s)$$

Rocks containing phosphorus compounds are quite common in the Earth's crust and phosphorus is an important element for the maintenance of life. Plants need it and so it is a necessary component of **fertilizers**, for example as **superphosphate**. Some **enzymes** used in **respiration** also contain the element. Animal bones contain phosphates.

Phosphorus was once used extensively in making the heads of **matches**, but has since been banned. Matches nowadays contain phosphorus sulphide. More than 90% of phosphorus produced today goes to make **phosphoric(V) acid**.

photochemical reactions Chemical reactions that are brought about by the action of electromagnetic radiation. In these reactions light provides the energy that makes the reactions work. The process of *photography* depends on photochemical reactions. When silver bromide on photographic film is exposed to light it decomposes:

$$2AgBr(s) \rightarrow 2Ag(s) + Br_2(g)$$

The black part of the photographic negative is thus a thin layer of silver. The negative is then processed to produce the final pictures.

Photosynthesis is the most important photochemical reaction. Our existence on this planet depends on it. Other examples are the reaction between hydrogen and chlorine which can be started off by a bright light (e.g. burning magnesium):

$$H_2(g) + Cl_2(g) \rightarrow 2HCl(g)$$

In hot areas of the world, such as California, where there are a lot of vehicle exhaust fumes, photochemical *smog* can be produced by the reaction of the exhaust fumes. This is an acrid haze that is very unpleasant.

The bleaching action of the sun is also a photochemical reaction, as is the conversion of cholesterol in the skin into **vitamin** D.

photodegradable Able to be broken down by sunlight. *Photodegradation* is an example of a **photochemical reaction**.

photography *See* **photochemical reactions**.

photosynthesis A photochemical process (*see* **photochemical reactions**) in which carbon dioxide and water are converted into carbohydrate and oxygen:

$$6CO_2(g) + 6H_2O(l) \rightarrow C_6H_{12}O_6(s) + 6O_2(g)$$

This occurs in the leaves of plants and is the means whereby plants obtain their food. It also releases oxygen back into the atmosphere.

The reaction is catalysed by **chlorophyll** and sunlight provides the energy for the reaction.

physical change A change to the physical **properties** of a substance which involves no alterations to its chemical properties.

physical chemistry The study of the physical **properties** of **elements** and **compounds** and the relationship between their chemical properties and physical properties.

pickling **1.** The treatment of metals with (usually warm dilute) acid in order to remove the surface layer of **oxide**, so that they can be painted or treated in some way. It is necessary to pickle steel if it has been allowed to cool down in the air.
2. The treatment of foods with **vinegar** in order to preserve them for eating at a later date. Chutneys, sauces and pickles are examples.

pig iron Another term for the **iron** produced in the **blast furnace**.

pigment A chemical that gives colour to a material. Pigments may be **organic** and be produced from **petroleum** and natural sources, or **inorganic** and be based on metal **oxides**. The white pigment, titanium oxide (TiO_2), makes up 70% of all pigments used.

pipette A piece of glassware used for measuring and transferring a fixed volume of liquid. The liquid is drawn up into the pipette through the jet by suction applied at s (*see diagram*). Liquid is taken up until the bottom of the meniscus is on the graduation.
   To remove the liquid from the pipette, the suction is released and the jet is touched against the side of the vessel into which the liquid flows. The pipette will then release the liquid, which will be exactly the correct volume. Note that a safety pipette filler should always be used to fill a pipette. *See also* **burette**.

graduation mark on the glass

pipette

jet

planar A molecule that has all its atoms in the same plane. In other words, it is flat. All **linear** and triatomic molecules must be planar. Examples include

$$H_2O \quad CO_2 \quad C_2H_2 \quad CO \quad HCl$$

Most carbon compounds are non-planar because of the **tetrahedral** arrangement of atoms around the carbon atom. Most alkenes are non-planar but **ethene** is an exception. (See figure overleaf.)

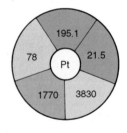

planar   *Ethene (left) is planar but propene (right) is not.*

plaster of Paris   A form of **gypsum** that has had half of the **water of crystallization** removed by heating. When water is added to plaster of Paris, a chemical reaction takes place to re-form gypsum.

$$CaSO_4.H_2O + H_2O \rightarrow CaSO_4.2H_2O$$

This is accompanied by a slight increase in volume which makes plaster of Paris ideal for use in moulds for making copies of artifacts.

plastic   **1.** Capable of being shaped or moulded by heat and pressure.
  2. A synthetic **polymer**. Examples are **poly(ethene)**, **poly(propene)**, **poly(carbonate)**, **polyamide (nylon)** etc. More than 30 different types of polymers exist and by combining them it is possible to produce a synthetic substance with specific properties. *See also* **thermosetting polymer**, **thermoplastic polymer**.

plasticizer   An **additive** that makes **polymers** and other materials more **flexible** and resilient (better able to withstand being hit).

platinum (Pt)   A **transition metal** that is used as a **catalyst** in the chemical industry. It is also used as a metal for jewellery and as inert **electrodes** in **electrolysis**. It is expensive because it is useful and rare. Platinum occurs in nature as the element.

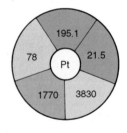

platinum

plutonium (Pu)   An artificial element. It is made in nuclear reactors from the 238 **isotope** of uranium by the absorption of a **neutron**:

$$^{238}_{92}U + ^{1}_{0}n \rightarrow ^{239}_{92}U \text{ (this is unstable)}$$
$$^{239}_{92}U \rightarrow ^{239}_{93}Np + ^{0}_{-1}e \text{ (beta particle)}$$

Neptunium-239 is also unstable:

$$^{239}_{93}Np \rightarrow ^{239}_{94}Pu + ^{0}_{-1}e \text{ (beta particle)}$$

Uranium-239 and neptunium-239 have a **half-life** of 20 minutes and two days respectively, and so soon decay. The half-life of plutonium-239 is more than 24,000 years and so is fairly stable.

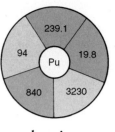

plutonium

Plutonium is used in bombs and as a fuel for nuclear reactors, such as fast breeder reactors. The cores of these reactors are surrounded by uranium-238 which is converted to plutonium-239 (in the way shown above) while the reactor is working. In this way it is possible to produce more fuel than is used in the reactor.

Plutonium is a toxic and dangerous chemical and must be handled with great care. Great controversy surrounds its use.

poison Any substance that is harmful if taken into the body and that will lead to injury or death. Poisons can be breathed in, e.g. hydrogen cyanide (HCN), carbon monoxide (CO) or lead fumes in the air; they may be eaten or drunk, e.g. arsenic(III) oxide, strychnine, or paraquat; or they may pass into the body through the skin, e.g. **benzene** and nerve gases. Radiation from nuclear **isotopes** can also be viewed as a poison.

polar Describing a molecule or a substance (especially a **solvent**) that contains **polar bonds**. Examples of polar solvents include water and **ethanol**.

polar bond A **covalent bond** in which the electrons spend most of their time closer to one atom in the bond rather than the other. As a result one atom has a slight negative charge while the other has a slight positive charge.

pollution The release into the environment of dangerous or dirty materials (*pollutants*). Pollution can occur in the air, in water and on land, when waste materials are not treated properly or when **toxic** materials escape from containers.

Examples of pollutants are:
(a) litter (paper, cans, glass) that is thrown away on the streets or in the countryside;
(b) exhaust gases (carbon monoxide, oxides of nitrogen) that enter the air from engines and factories;
(c) **fertilizers** (phosphates, nitrates) from farms, that are allowed to enter streams and rivers;
(d) **petroleum** that escapes from tankers and is washed up onto a beach.
(e) toxic chemicals that are dumped in places from which they can escape into the ground water.

See also **acid rain, global warming**.

poly- A prefix meaning 'many', used in the naming of chemicals to indicate that a compound is a **polymer**. Thus a **polyamide** is a polymer containing many amide groups.

polyamide A synthetic **polymer** that contains the *amide* group of **atoms** (–CONH–). Some such compounds are known by the name **nylon**. Polyamides are **condensation** polymers formed by the reaction of an amine and an acid.

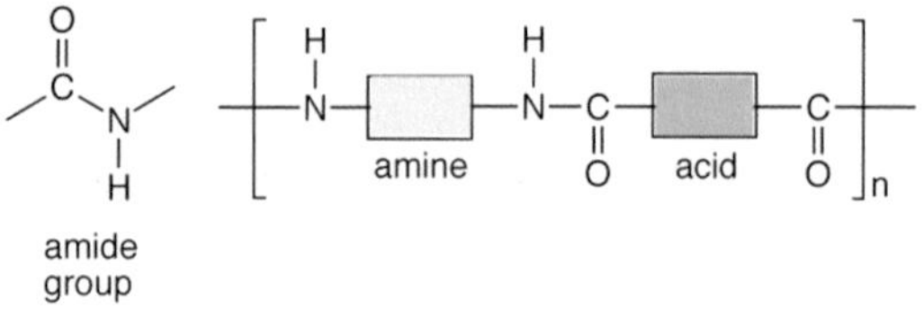

amide group

polyamide

poly(carbonates) Transparent **polymers** that have resistance to heat and other chemicals. They are used in safety spectacles, lenses and babies' bottles.

poly(chloroethene) (commonly called *polyvinylchloride* (PVC)) A **polymer** that is made from chloroethene ($CH_2=CHCl$). PVC has a wide range of uses: in vinyl flooring, food containers, fibres, guttering, bottles, windows and cables.

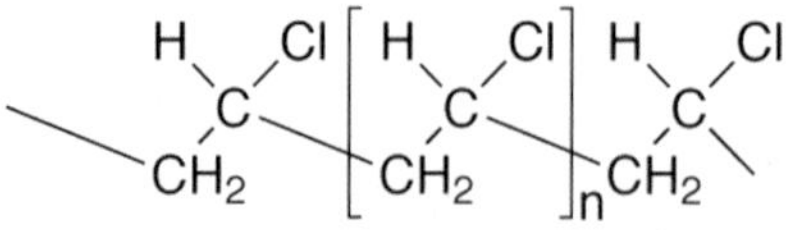

poly(chloroethene)

polyesters Synthetic **polymers** that contain an **ester** group of atoms. They are **condensation** polymers whose **monomers** are an acid and an **alcohol**. Polyesters are mainly used to produce **fibres** (80%), film (7%) and packaging materials (10%).

Polyester fibre is used to produce clothing, sometimes blended with other fibres such as wool and cotton. Because it is a good insulator, it is used as a filling for anoraks and duvets. The fibres are also used in car tyres and hoses to give strength. *Terylene* is a polyester widely used in the clothing and household furnishing industries and is often mixed with natural fibres.

Polyester film and sheeting is used for videotapes, photographic films and insulating tape. Modern developments include its use in bottles to hold carbonated drinks. (*See* figure opposite.)

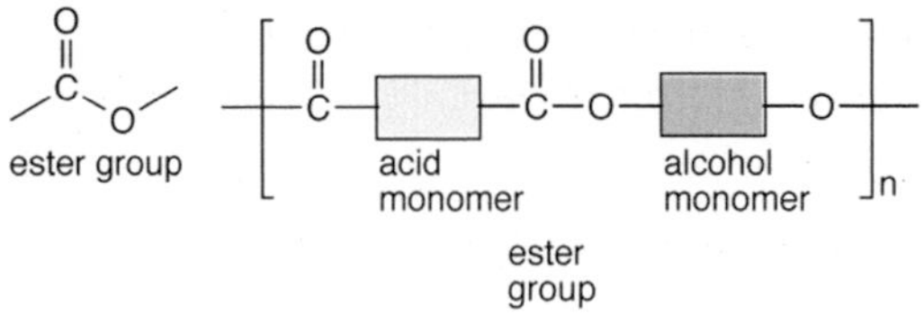

polyesters

poly(ethene)   A **polymer** that is made from **ethene** and is usually known as *polythene*.

Poly(ethene) is an addition **polymer**. A poly(ethene) molecule will contain a very large number of ethene groups (in the diagram $n > 50,000$). The polymer is a saturated alkane (*see* **saturated compounds**) and so is very unreactive.

Poly(ethene) has wide uses. It is produced in a *low density* form (LDPE) and a *high density* form (HDPE). LDPE is made into film for packaging, and HDPE is blow-moulded to produce containers such as bottles for washing-up liquid, and pipes and gutters. Both forms can be used to make food boxes and bowls by the injection moulding process.

The **monomer**, ethene, is produced by **cracking** the products of **petroleum** refining, e.g. **naphtha** and **ethane**.

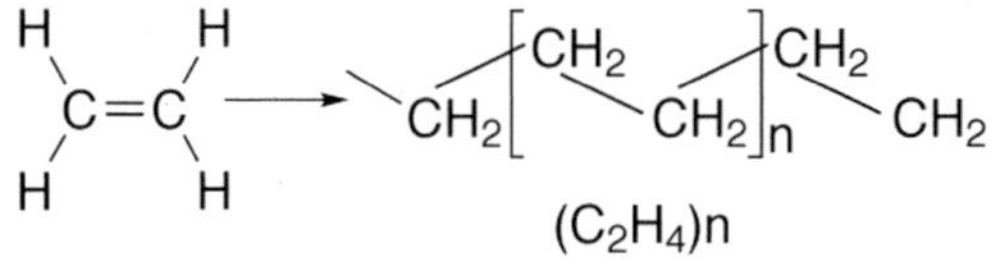

$(C_2H_4)n$

poly(ethene)

polymer   A large **molecule** in which one or more group of atoms (or **monomers**) are repeated. For example:

$$-x-x-x-x-x-x-x-x$$

or

$$-x-y-x-y-x-y-x-y$$

Polymers can be naturally occurring or **synthetic**. **Starch** and **cellulose** are natural, whilst **nylon**, polyesters, **poly(choroethene)** and **poly(propene)** are synthetic.

Polymers are made by reacting monomer molecules together usually in the presence of a **catalyst**. The process of making a polymer from its constituent monomers is known as *polymerization*.

In *addition polymerization,* one kind of unsaturated molecule (*see* **unsaturated compound**) links together to form a long chain of atoms (now joined by **single bonds**). *See* **poly(ethene)**. Because the polymer is a saturated molecule (*see* **saturated compound**) it is usually fairly unreactive and this gives the polymer useful properties.

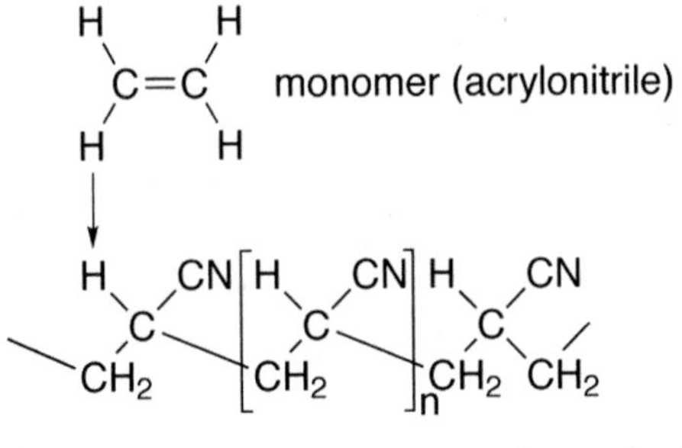

polymer *Addition polymerization used to make filaments which are manufactured into fabrics such as Acrilan.*

In *condensation polymerization* (*see* **condensation**), two molecules (the monomers) condense together into long chains. A small molecule is eliminated during the reaction. The monomers are difunctional molecules, i.e. each molecule possesses two functional groups. **Nylon** and polyester are good examples of condensation polymers. Nylon is made by reacting an acid and an amine.

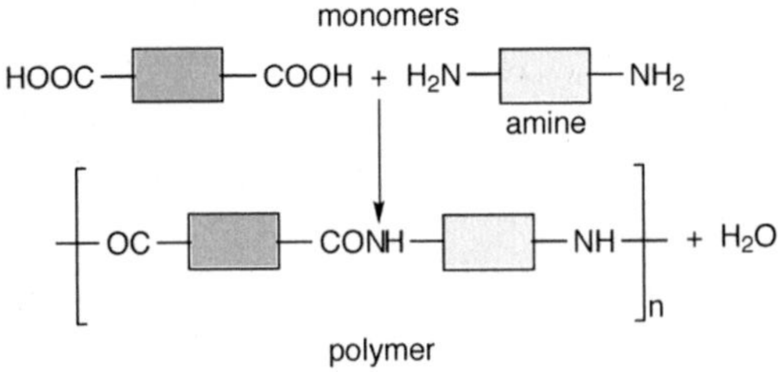

polymer *Condensation polymerization showing the formation of a polyamide.*

The source of most of the raw materials used in making polymers is **petroleum**.

See also the entries for individual polymers.

polymerization   The reaction of **monomers** to form a **polymer**.

poly(methylmethacrylate)   An addition **polymer** that is transparent. It is stronger than **glass** but easier to scratch and is used widely as a substitute for glass. It is also used to produce baths, car light clusters and aircraft windows. *Perspex* is a common trade name.

polypeptide A molecule that consists of up to 50 **amino acids** linked together with **peptide** links. If there are more than 50 amino acids the molecule is termed a **protein**. Polypeptides are important molecules in the body, acting, for example, as **enzymes**.

poly(phenylethene) or polystyrene A **polymer** based on the **alkene** phenylethene. It is an addition polymer and is used widely in the form of sheets, mainly in an expanded form (*expanded polystyrene*). Here air is blown into the polymer and the result is a white granular solid that has excellent insulation properties. It is used for cups, ceiling tiles and other items which need to be poor conductors of heat. It is also a useful packaging material.

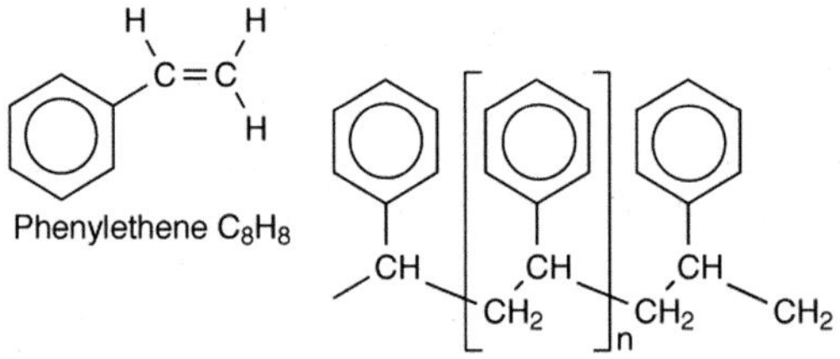

Phenylethene C_8H_8

poly(phenylethene)

poly(propene) or polypropylene A **polymer** similar to **poly(ethene)** in that it is an addition polymer produced from an **alkene**. In poly(propene), the alkene is **propene** (C_3H_6), which is obtained from **petroleum**. The polymer is versatile and useful. It is strong and hard wearing and finds use as carpet fibres and material for laboratory flasks and beakers. When injection-moulded, poly(propene) can be made into a wide range of everyday objects such as washing-up bowls and car bumpers. It is very unreactive.

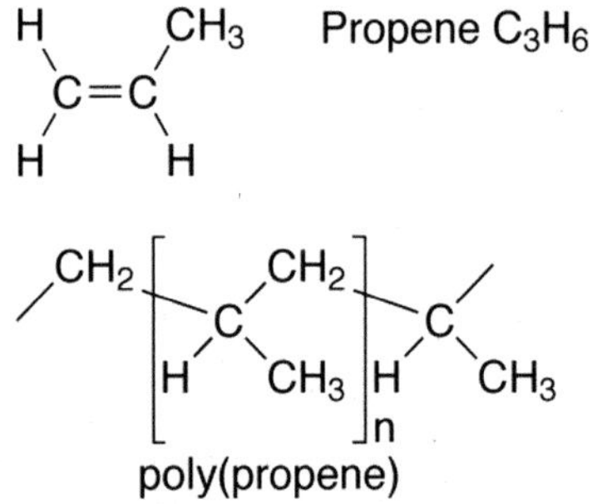

Propene C_3H_6

poly(propene)

poly(propene)

polysaccharide A **polymer** made up of **sugar** molecules such as **glucose**. There are three common naturally occurring examples, **starch**, **cellulose** and glycogen, all of which are made up of chains of glucose molecules.

Polysaccharides can be broken down by **hydrolysis** into simple sugars such as glucose or into **disaccharides** such as **maltose**. Hydrolysis is brought about by **enzymes** in plants and animals but can also be done by the use of **acids**.

Polysaccharides are valuable foods; an example is starch in the form of potatoes or rice; cellulose is found in all plants and glycogen is the form in which excess **carbohydrate** is stored in animals.

polystyrene *See* **poly(phenylethene)**.

poly(tetrafluoroethene) (PTFE) A **polymer** that is very inert. It is used for the non-stick surface of saucepans, for example as the brands Teflon and Fluon. It is also used for bearings and for replacement joints in the body because it has a low coefficient of friction.

$$\diagdown CF_2 \diagup \left[\diagdown CF_2 \diagup \right]_n \diagdown CF_2 \diagup$$

poly(tetrafluoroethene)

polythene The older term for **poly(ethene)**, which is still used in industry and shops.

polyurethanes *See* **urethanes**.

porous (used of a material) Containing minute passages (*pores*) through which fluids can pass.

positive electrode *See* **anode**.

potassium (K) A group I metal. It is a grey, very soft element that is very reactive. It is easily cut with a knife, revealing a silvery surface which tarnishes immediately. It is a powerful **reducing agent**, giving rise to the potassium ion (K^+):

$$2K(s) + 2H_2O(l) \rightarrow 2KOH(aq) + H_2(g)$$

It is stored under oil because of its reactivity towards air and water. The metal **ion** gives a lilac **flame test**. Potassium is obtained by the **electrolysis** of molten potassium chloride (KCl).

potassium compounds   Colourless, crystalline compounds that are very soluble in water. Potassium **salts** play an important role in the body.

Potassium iodide KI	The solution is a useful solvent for iodine forming KI_3(aq). This is used to test for starch.
Potassium nitrate KNO_3	Saltpetre. Used in food preservatives and in explosives.
Potassium manganate(VII) $KMnO_4$	A vivid purple crystalline compound which is a good oxidizing agent.

potassium compounds

precipitate (ppt)   The insoluble substance formed on mixing the two solutions in a **double decomposition** reaction. For example:

$$Pb(NO_3)_2(aq) + 2NaCl(aq) \rightarrow PbCl_2(s) + NaNO_3(aq)$$

Lead(II) chloride is the precipitate.

pressure   A measure of the force pressing onto an object's surface. The **SI unit** of pressure is the **pascal**. Other units used are **atmospheres** and **mmHg**.
  The pressure that a gas exerts upon its container is caused by molecules striking the container's walls. The pressure of a gas depends upon its volume and its temperature. *See also* **Boyle's law**, **Charles' law**, **gas laws**, **rate of reaction**.

product   *See* **reaction**.

propane   An **alkane** that is obtained from **petroleum**. It is chiefly used as a **fuel**:

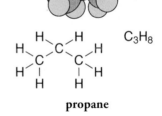

$$C_3H_8(g) + 5O_2(g) \rightarrow 3CO_2(g) + 4H_2O(g)$$

Although a gas at room temperature, it is easily liquified and it is sold as bottled gas, which is mixed with **butane**.

propane

propene   A **hydrocarbon** gas with the formula C_3H_6. Propene is an **alkene** and is the **monomer** of an important **polymer**, **poly(propene)**.

properties   The characteristic ways in which a substance behaves (reacts), which make it what it is and differentiates it from other substances.
  It is usual to classify properties as either physical or chemical. *Chemical properties* are concerned with the substance's reactions.

Physical properties	Chemical properties
Colour	Whether the substance is a metal or non-metal
Density	Gives acidic or basic oxides
Physical state	Has more than one valency
Boiling point	Reacts with acids
Melting point	Is an oxidizing or reducing agent
Crystal form	
Solubility	
Hardness	

properties *Physical and chemical properties*

proteins   Large, organic compounds made up of **chains** of **amino acids**. The amino acids are joined by **peptide** links. Proteins are widespread in animal bodies: skin, hair, nail, wool, gristle, muscle are all formed of protein.

Proteins form a vital part of our diet. Good sources are meat, nuts, fish, eggs, milk, cheese, beans, bread. The **enzymes** that play so important a role in living organisms are made up of protein molecules. Animals obtain the food which they need in order to make proteins in their bodies by eating other animals or plants. Plants, by contrast, make their own proteins from the nitrogen-containing compounds in the soil. *See also* **fertilizer**.

proton   A positively charged **subatomic particle** found in the **nucleus** of the **atom**. The number of protons in an atom (the *proton number*) is the same as the **atomic number**. The proton number is the same for all the **isotopes** of an element.

PTFE   *See* **poly(tetrafluoroethene)**.

pure   Containing only one element or compound. For example, sodium chloride crystals can be obtained in a pure state, but table salt is impure because it is mixed with other substances.

purification   The process of removing impurities to make a substance **pure**.

PVC (polyvinylchloride)   *See* **poly(chloroethene)**.

pyrites   A sulphide mineral. Its formula is FeS_2 and the mineral has a characteristic brass colour.

pyrolysis   The decomposition of a substance by the action of heat, for example:

$$CaCO_3(s) \rightarrow CaO(s) + CO_2(g)$$

Q

qualitative (used of a statement or **analysis**) Concerned with composition, not with amounts. For example, 'Water is a compound of hydrogen and **oxygen**' is a qualitative statement.

quantitative (used of a statement or **analysis**) Concerned with amounts. For example, 'Water consists of two atoms of hydrogen and one of oxygen' is a quantitative statement.

quartz The transparent mineral form of **silica** (silicon dioxide SiO_2) that is used in optical instruments such as watches, clocks, microscopes and spectroscopes. More **ultraviolet radiation** passes through quartz than through **glass**.

quicklime The old name for calcium oxide (CaO).

R

radical (in inorganic chemistry) The atom or group of atoms present in a compound that is responsible for the characteristic properties of that compound. Examples are the sulphate group $-SO_4^{2-}$ and the carbonate group $-CO_3^{2-}$.

radioactive *See* **radioactivity**.

radioactivity The spontaneous disintegration of the **nucleus** of an **atom** accompanied by the emission of electromagnetic radiation (**gamma rays**) or particles (**alpha particles** or **beta particles**).

Not all elements have radioactive **isotopes**, though many can now be created artificially. Some radioactivity occurs naturally, e.g. isotopes found within rocks. Granite usually has quite a large concentration. Living things contain radioactive carbon and this can be used for finding the age of their remains (*see* **carbon dating**). *See also* **nuclear reactions**, **half-life**.

radiocarbon dating *See* **carbon dating**.

radon (Rn) An element that shows radioactivity. It is a member of group 0 of the **periodic table**, the **noble gases**.

RAM *See* **relative atomic mass**.

rate of reaction The rate (or speed) at which a reaction occurs. This depends on several factors:
(a) *Temperature:* The higher the temperature, the faster the reaction is because the particles are moving with greater energies.
(b) *Particle size:* The smaller the particles involved, the greater the surface area where the reaction can take place and the faster is the reaction.
(c) *Concentration:* The more concentrated a solution (or the higher the pressure of a gas) the faster a reaction will occur. This is because there are more particles in a given volume to react.
(d) *Catalysts:* These change the rate by providing an alternative reaction pathway along which the reaction can occur.

raw materials Materials used as the starting point in the production of chemicals or manufactured goods. The major raw materials used in the UK chemical industry are **hydrocarbons**, air, metallic ores, minerals and water.

For example, the raw material in the production of chlorine is **rock salt**, which is dissolved in water and electrolysed (*see* **electrolysis**).

In the production of ammonia, the raw materials used in the UK are **natural gas** and air.

rayon   One of the first **synthetic** fibres (although it is made from a natural material, **cellulose** (from wood)). The cellulose fibres in wood are too short to spin into yarn. The cellulose is therefore dissolved in sodium hydroxide/carbon disulphide solution and squirted into dilute sulphuric acid. Here the cellulose is reformed as long filaments which are then either twisted to form a yarn or chopped and then spun into a yarn to be ready for weaving. Rayon is used to make clothes fabrics and curtaining.

reactant   *See* **reaction.**

reaction   A change in which one or more chemical elements or compounds (the *reactants*) form one or more new compounds (the *products*). The products will have different properties from the reactants. Reactions can be accompanied by heat, light, sound and colour changes. *See also* **equilibrium, reversible reaction.**

reaction types   Types of chemical **reaction**. *See* **addition reaction, dehydration, displacement reaction, dissociation, hydration, hydrolysis, neutralization, polymerization, redox, substitution reaction.**

reactivity series   A series of **elements** arranged in order of their chemical reactivity. The series shown here is a short one containing only a few elements but it shows the general principle. **Oxides** will be reduced by elements above them but not by those below them:

Potassium
Sodium
Magnesium
Aluminium
Zinc
Iron
Copper
Silver

reactivity series

$$CuO(s) + Mg(s) \rightarrow MgO(s) + Cu(s)$$

$$MgO(s) + Cu(s) - \text{no reaction}$$

Metal **ions** will be reduced by elements above them but not by those below them:

$$2Ag^+(aq) + Zn(s) \rightarrow Zn^{2+}(aq) + 2Ag(s)$$

$$Zn^{2+}(aq) + Ag(s) - \text{no reaction}$$

The higher you go up the series, the more vigorous is the reaction with water or acids. Potassium reacts violently with cold water, magnesium reacts very slowly with cold water but vigorously with steam. Copper reacts with neither. The reactivity series includes only metals whereas the **electrochemical series** also includes **non-metals.**

recrystallization   A process for the **purification** of impure crystalline substances. They are dissolved in a solvent, filtered to remove the impurities, and allowed to crystallize again.

recycling   The re-use of materials. There is an increasing emphasis on recycling as resources become scarce and expensive. Recycling saves energy costs and usually makes a country less dependent upon imported goods. Paper, glass, steel and aluminium are examples of materials that can be recycled. Plant material can be recycled as fertilizer by making it into compost.

redox   A chemical reaction involving both **reduction** and **oxidation**. Any reaction that involves an oxidation must also involve a reduction.

$$\text{Mg(s)} + \text{CuO(s)} \rightarrow \text{MgO(s)} + \text{Cu(s)}$$

with *oxidation* shown over the Mg→MgO and *reduction* over the CuO→Cu.

$$\text{Cu}^{2+}\text{(aq)} + \text{Zn(s)} \rightarrow \text{Cu(s)} + \text{Zn}^{2+}\text{(aq)}$$

with *oxidation* shown over the Zn→Zn²⁺ and *reduction* over the Cu²⁺→Cu.

redox *Examples of reactions.*

reducing agent   A chemical that brings about the **reduction** of a substance. In the reaction the reducing agent is itself always oxidized (*see* **oxidation**). Common reducing agents are hydrogen, carbon, carbon monoxide, sulphur dioxide.
    Some examples of reducing agents at work are:

$$\text{Fe}_2\text{O}_3\text{(s)} + 3\text{CO(g)} \rightarrow 2\text{Fe(l)} + 3\text{CO}_2\text{(g)}$$

$$\text{CuO(s)} + \text{H}_2\text{(g)} \rightarrow \text{Cu(s)} + \text{H}_2\text{O(g)}$$

$$\text{ZnO(s)} + \text{C(s)} \rightarrow \text{Zn(s)} + \text{CO(g)}$$

reduction   A chemical reaction in which one of the following occurs to a substance:
(a) it loses oxygen $\text{PbO(s)} + \text{C(s)} \rightarrow \text{Pb(s)} + \text{CO}$
(b) it gains hydrogen $\text{Cl}_2\text{(g)} + \text{H}_2\text{(g)} \rightarrow 2\text{HCl(g)}$
(c) it gains electrons $\text{Na}^+\text{(l)} + \text{e}^- \rightarrow \text{Na(l)}$
See also **oxidation**, **reducing agent**, **redox**.

refining   A process for either the removal of impurities from a substance such as a metal, or the removal of components from a mixture such as petroleum.
    When a metal has been extracted from its ore it is often not in a pure state. In some cases this might be acceptable, e.g. pig iron contain impurities. When a purer product is required, the metal is refined until the right purity is achieved.

Petroleum is a complex mixture of **hydrocarbons**. To make it more useful it is first of all split up into separate parts, known as *fractions*, by means of **fractional distillation**. The fractions are also mixtures but are not so complicated. They contain compounds whose boiling points are within a certain range.

Conditions vary from refinery to refinery and depend on what products are required. The major product might be **petrol** or it might be petrochemical feedstock which is then turned into **alkenes**, etc., for the production of polymers. *See also* **cracking**, **reforming**.

reforming   An important process that is used in **petroleum** refineries and chemical plants. Molecules are taken and altered into more useful products. There is no change in the size of the molecule. One example is the conversion of unbranched **alkanes** into branched ones (*see diagram*), which give **petrol** a higher **octane rating**:

octane

2-methylheptane

Another example is the production of **aromatic** compounds from **alkenes**:

$$C_7H_{14} \rightarrow C_7H_8 + 3H_2$$
heptane   methylbenzene hydrogen

refractory materials   Materials that neither melt nor change in any way below 1500 °C and which can be used as furnace linings. Examples include **china clay**, silica, alumina and **bauxite**.

relative atomic mass (A_r)   The **mass** of an 'average atom' of an element compared with the mass of $^1/_{12}$ of an atom of the $^{12}_6C$ carbon **isotope**, which is given the value of exactly 12. *See also* **relative molecular mass**.

relative formula mass   A term that can be used in place of **relative molecular mass** when describing **giant structures** as well as molecules.

relative molecular mass (M_r)   The mass of a molecule compared with the mass of $^1/_{12}$ of an atom of the $^{12}_6C$ carbon isotope. It can be calculated by adding together the **relative atomic masses (A_r)** of each atom within the molecule. For example:

M_r carbon monoxide (CO) = 12 + 16 = 28

M_r water (H_2O) = (2 × 1) + 16 = 18

residue   The waste product remaining after any chemical process is complete.

respiration   The process by which **energy** is obtained by plants and animals. In animals, such as humans, food (such as energy-rich **carbohydrates**) is eaten and broken down in the body. Oxygen is breathed in by taking air into the lungs. The oxygen is carried round the body in the blood to the cells. Here reactions occur between the oxygen and the molecules of broken-down food (typically **glucose**), releasing chemical energy from the food-molecules. Carbon dioxide is also produced and this is breathed out. The waste products from the food are excreted from the body.

   The energy released from food by respiration is used for heat, movement, growth and all other body functions.

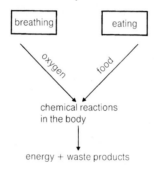

respiration

reversible reactions   Reactions that can go in either direction, depending on the conditions that exist. For example:

$$Fe_2O_3(s) + 3H_2(g) \rightleftharpoons 2Fe(s) + 3H_2 + 3H_2O(l)$$

$$ICl(l) + Cl_2(g) \rightleftharpoons ICl_3(s)$$

Steam can react with hot iron, or hydrogen can reduce iron(III) oxide. In both cases, an **equilibrium** will exist with all four substances present unless the products are removed.

rhombic sulphur   *See* **sulphur**.

RMM   *See* **relative molecular mass**.

rock cycle **or** geochemical cycle   The series of processes over long periods of time (typically millions of years) by which rocks on Earth are changed from one kind to another. They are broken down on the surface by physical or chemical means, and altered or even melted within the crust. Eventually the new rock is brought back to the surface by great movements of the Earth or by volcanoes, and the cycle starts again. *See also* **Earth's structure, igneous rock, metamorphic rock, sedimentary rock.**

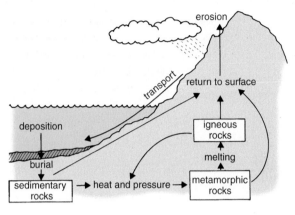

rock cycle

rock salt Impure crystalline deposits of sodium chloride (*see* **sodium compounds**). The solid is purified before use.

room temperature A temperature of between 15 °C and 25 °C. It is not a fixed temperature but a range.

rubber A natural **polymer**. It is a **hydrocarbon**. Today most 'rubber' is made from butadiene

$$CH_2=CH–CH=CH_2$$

by polymerization (*see* **polymer**). Rubber is elastic and a good **insulator**. Its uses are widespread, from tyres and gloves to waterproofing.

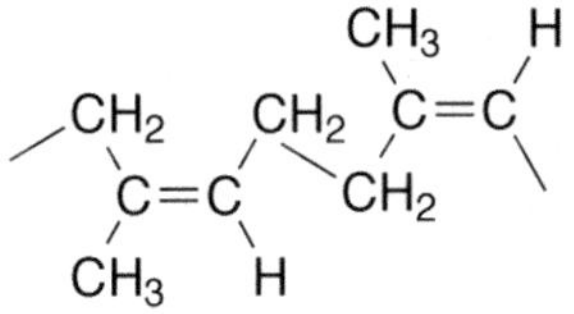

rubber

rust The reddish-brown product of **corrosion (oxidation)** of iron that has been exposed to air and water. It is hydrated iron(III) oxide ($Fe_2O_3 \cdot xH_2O$). Rusting is of great economic importance. It is most commonly prevented by coating the iron with paint, plastic or another metal. *See also* **electroplating, galvanizing.**

S

sacrificial anode A reactive metal that is attached to another metal or alloy that needs protection from **oxidation**. For example, bars of zinc are welded to undersea pipelines so that the more reactive zinc will oxidize in preference to the less reactive iron. In this way the steel of the pipeline is protected. **Galvanizing** is another example of this process. *See also* **anode**.

salt A compound formed when the hydrogen of an **acid** is totally or partially replaced by a metal, e.g.:

$$Zn(s) + HCl(aq) \rightarrow ZnCl_2(aq) + H_2(g)$$

When an acid reacts with a **base** the products are a salt and water only:

$$NaOH(aq) + HNO_3(aq) \rightarrow NaNO_3(aq) + H_2O(l)$$

Salts can also be made by direct combination of two elements:

$$2Na(s) + Cl_2(g) \rightarrow 2NaCl(s)$$

With a **dibasic acid**, if only one hydrogen atom is replaced, the result is an **acid salt**.

 'Salt' is the everyday word for sodium chloride (common salt); *see* **sodium compounds**.

salt bridge A pathway containing an **electrolyte** that connects two parts of a **cell**.

sand A mixture of fine fragments of rock. Sand is widely found on Earth and is extensively used as a building material. It contains a high proportion of **silica** (SiO_2).

sandstone A common **sedimentary** rock consisting of **sand** grains bound together by a **silica**-rich or **carbonate** 'cement'.

saponification The **hydrolysis** of an **ester** when alkaline conditions are used. The breakdown of natural fats to produce **soap** is an example of saponification. Here sodium hydroxide solution is used to produce an **alcohol** and the sodium salt of the **carboxylic acid** (soap). *See also* **detergent**.

saturated compound An organic compound containing only **single bonds**. Compounds that contain **double bonds** or **triple bonds** are said to be **unsaturated compounds**. Examples include ethane and butan-1-ol. *See also* **hydrocarbon**.

saturated solution   A **solution** that contains the maximum amount of **solute** at a given temperature in the presence of surplus solute. The amount of solute needed to form a saturated solution depends on the temperature. The only way of putting more solute into a saturated solution is by increasing the temperature. *See also* **supersaturated solution**.

sedimentary rock   **Rock** formed from *sediments* – solid materials – that have been laid down in water (chiefly the sea) over millions of years. As the layers of materials build up, the pressure increases and the particles are compressed to form rock. Most of the Earth's crust is covered by a thin layer of sedimentary rock.

The original sediments are formed from materials such as **sand**, gravel and mud produced by the action of weather on existing rocks. Examples of sedimentary rocks are **sandstone**, **limestone** and **chalk**. *See also* **rock cycle**.

semiconductor   An electrical **conductor** with unusual properties. As the temperature increases or the substance contains greater amounts of impurity, the resistance of the material decreases. In practice, when **crystals** of semiconductors are grown, controlled amounts of impurity are added to obtain exactly the properties which are required.

Substances that have semiconduction properties can be elements or compounds but usually involve **metalloids**. Examples include gallium, germanium, arsenic and silicon.

separation of mixtures   The separating-out of one or more of the substances that make up a **mixture.** A mixture can be separated if the substances in it have different physical properties. Which technique is used depends on the mixture. Techniques include filtration (*see* **filter**), **fractional distillation**, **chromatography**, decantation.

sewage treatment   A technique for making waste water, from industry or the home, fit again for use. Several processes are involved:
(i) the material is passed through a screen to remove large solid objects, and sand and grit are allowed to settle out;
(ii) it is then passed through settling tanks where small particles sink to form a sludge; this sludge is treated with bacteria, and the **methane** thus formed can be collected and used. The treated sludge can be used as a **fertilizer**;

(iii) the remaining liquid is then treated in one of two ways; either:
(a) it is sprayed over a bed of coke which contains bacteria; these digest any remaining organic material; or
(b) it is mixed with microorganisms and churned up by having air passed through it; again, the organic material is digested.
The clean water is then ready for discharge into rivers.

shell *See* **electronic configuration**.

shift reaction A process in which carbon monoxide reacts with steam to form carbon dioxide and hydrogen:

$$CO(g) + H_2O(g) \rightarrow CO_2(g) + H_2(g)$$

See **steam reformation, synthesis gas**.

silica A widely found mineral form of silicon dioxide (SiO_2) which is used as an ore of silicon. Some silica is found as coloured crystals, e.g. amethyst (containing the Fe^{3+} **ion**); other forms of silica include flint, which is an **amorphous** form. **Quartz** is the purest form of silica. **Sand** is tiny fragments of silica with various impurities.

silicon (Si) A group IV **metalloid**. It is a very abundant element, the second most abundant in the Earth's crust (27.7%), being a part of the chemical composition of many rocks. It is a **semiconductor** and is the heart of microelectronic technology in the form of the silicon chip. Silicon is found naturally in the form of metal silicates and **silica** (silicon dioxide).

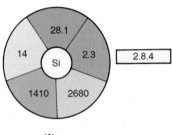

silicon

silver (Ag) A **transition metal**. It is used for jewellery and decorative purposes (both as the solid metal and as a thin coating on another metal: silver plate) and has been used for coinage. It is often alloyed with copper to give strength. It is an excellent conductor of electricity and heat. When exposed to the air it slowly becomes covered in a black film of silver sulphide.

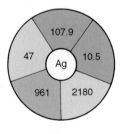

silver

Silver is low in the **reactivity series**. Its **oxide** is unstable:

$$\text{Ag}_2\text{O(s)} \xrightarrow{\text{heat}} 2\text{Ag(s)} + \text{O}_2\text{(g)}$$

Its **halides** are also unstable. This property is used in photography:

$$2\text{AgBr(s)} \xrightarrow{\text{light}} 2\text{Ag(s)} + \text{Br}_2\text{(g)}$$

See also **photochemical reactions**.

single bond   A **covalent bond** that is made up of a shared pair of **electrons**.

SI units   An international system of **units** based on the metric system of measurement. There are seven basic units and the system is used in scientific and technological work throughout the world.

Quantity	Unit	Symbol
Length	metre	m
Mass	kilogram	kg
Time	second	s
Current	ampere	A
Temperature	kelvin	K
Amount of substance	mole	mol
Luminous intensity	candela	cd

SI units

slag   A mixture of molten **oxides** produced during the **smelting** and **refining** of **ores**. Slag plays an important part in these processes because it dissolves impurities from the ores and keeps them separate from the liquid metal. To help this process, materials are added which reduce the melting point of the oxide mixture. Limestone is a good example. Examples of impurities that are removed during **steel manufacture** include carbon, phosphorus and sulphur. Solid slags are used for a variety of purposes which include cement production. *See also* **blast furnace**.

slaked lime   The common name for calcium hydroxide (Ca(OH)_2). It can be produced by adding water to **lime**, a process known as *slaking*:

$$\text{CaO(s)} + \text{H}_2\text{O(l)} \rightarrow \text{Ca(OH)}_2\text{(s)}$$

A solution of slaked lime in water is known as **limewater**. Slaked lime is used in agriculture and in making **mortar**. *See also* **calcium compounds**, **limestone**.

smelting A process of extracting metals that involves melting the **ore**.

smog A polluting mixture of smoke and fog. Photochemical smog is a mixture of dangerous compounds produced by **photochemical reactions**.

smoke A **colloidal dispersion** of solid particles in air.

soap or **soapy detergent** A cleaning agent that is made by the action of an **alkali**, such as sodium hydroxide solution, on naturally occurring esters. *See also* **saponification**.

$$H-\underset{\underset{H}{|}}{\overset{\overset{H}{|}}{C}}-\underset{\underset{H}{|}}{\overset{\overset{H}{|}}{C}}-\cdots-\underset{\underset{H}{|}}{\overset{\overset{H}{|}}{C}}-\overset{\overset{O}{\|}}{C}-O^-Na^+$$

soap *A soapy detergent molecule.*

The **hydrocarbon** end of the molecule is hydrophobic (water-hating). The other end (the carboxylate end) of the molecule is **hydrophilic**.

Soap is made from a mixture of animal fat and vegetable oils such as coconut oil. An **antioxidant** is usually added to prevent the soap from producing 'off' smells. Soap is **biodegradable**. *See also* **detergent**.

soapless detergent *See* **detergent**.

soapy detergent *See* **soap**.

sodium (Na) A soft, grey metal that is in group I of the **periodic table**. It is easily cut with a knife revealing a silvery surface that rapidly tarnishes. It is stored under oil because of its reactivity towards air and water. It is very reactive towards water and non-metals, e.g:

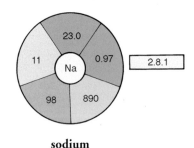

sodium

$$Na(s) + 2H_2O(l) \rightarrow NaOH(aq) + H_2(g)$$

$$2Na(s) + Cl_2(g) \rightarrow 2NaCl(s)$$

The metal **ion** gives a yellow/orange **flame test**. Sodium is extracted from molten sodium chloride by **electrolysis**:

$$2NaCl(l) \rightarrow 2Na(l) + Cl_2(g)$$

The sodium chloride is obtained by mining **rock salt**. Sodium metal is used as a coolant in fastbreeder nuclear reactors (*see* **nuclear reaction**) and in the manufacture of the **petrol** additive tetraethyl lead ($Pb(C_2H_5)_4$), which is used to raise the octane rating of fuels. Sodium is also used to produce **biocides**, pharmaceuticals and **catalysts**, and to extract metals, e.g. titanium:

$$TiCl_4(l) + 4Na(l) \rightarrow Ti(s) + 4NaCl(l)$$

Sodium ions (Na^+) are an important constituent of the fluids in animal tissues.

sodium compounds

Sodium carbonate Na_2CO_3	Compound made in the Solvay process and an important chemical. One of its many uses is in the manufacture of glass. Unlike most carbonates it is soluble in water. It is not decomposed by heat, its aqueous solution is alkaline.
Sodium chloride NaCl	Known as common salt. Obtained from rock salt and is used to make sodium hydroxide, sodium metal and chlorine. It is used for flavouring and preserving foods. *See also* **Solvay process**.
Sodium hydrogencarbonate $NaHCO_3$	Used in baking powder. Its is decomposed by the action of heat or acids, releasing carbon dioxide gas which makes cakes 'rise'. Also used in fire extinguishers and in anti-indigestion powders.
Sodium hydroxide NaOH	Produced by the electrolysis of brine and has wide uses in industry, e.g. making soap, fibres such as rayon and paper. It is a caustic alkali. Solutions have pH>10.

soft water *See* **hard water**.

sol *See* **colloidal dispersion**.

solder A fusible **alloy** used for bonding metals. It is usually an alloy of tin and lead in various proportions, depending upon the use.

solid A substance whose atoms or molecules are fixed in positions and do not have the freedom of movement found in a liquid or a gas. Atoms and molecules are held in a **lattice** by bonds. It is only when these bonds are broken that the atoms and molecules are able to move and the solid melts (*see* **melting**).

As **heat energy** is put into the solid lattice the atoms or molecules acquire enough energy to break the bonds. Although the atoms or molecules within the lattice do not move from place to place they do vibrate.

soluble (used of a substance) Able to **dissolve** in a **solvent**. The extent to which any **solute** dissolves depends on the solvent and the temperature. At any given temperature there is a maximum amount of solute which can dissolve in a fixed volume of solvent. This produces a **saturated solution**. *See also* **solubility**.

solubility   The extent to which a **solute** will **dissolve** in a **solvent**. An **ionic** substance will have a higher solubility in a **polar** solvent than a covalently bonded solute, for example copper(II) sulphate is much more soluble in water than is methane. The units used are **moles** of solute in 100 g of solvent at a stated temperature, although other units, such as **mole dm**$^{-3}$, g/100 g, are also used.

solute   Any substance that **dissolves** in a **solvent** to produce a **solution**. For example, when copper(II) sulphate dissolves in water to produce a solution, copper(II) sulphate is the solute.

solution   The liquid mixture resulting when a solid, liquid or gas **dissolves** in a **solvent**. In a solution, the particles of the **solute** (its atoms or molecules) are spread throughout the body of the solvent and are not visible. *See also* **colloidal dispersion**, **suspension**.

Solvay process   The process by which **brine** and calcium carbonate (in the form of **limestone**) are converted into the valuable alkali, sodium carbonate. The overall process can be represented by the equation:

$$2NaCl(aq) + CaCO_3(s) \rightarrow CaCl_2(aq) + Na_2CO_3(aq)$$

Because both sodium carbonate and calcium chloride are soluble in water, the reaction shown cannot be carried out directly. It does work, however, by a series of steps:
(i) limestone is heated: $CaCO_3(s) \rightarrow CaO(s) + CO_2(g)$

(ii) ammonia is dissolved in brine ($NaCl + NH_3$);
(iii) carbon dioxide (produced from the limestone) is passed into the brine/ammonia solution. The following reaction occurs:

$$NaCl(aq) + CO_2(g) + NH_3(g) + H_2O(l) \rightarrow NaHCO_3(s) + NH_4Cl(aq)$$

Sodium hydrogencarbonate is precipitated because of its low solubility. It can be removed from the mixture;

(iv) sodium carbonate is produced by heating the hydrogencarbonate:

$$2NaHCO_3(s) \rightarrow Na_2CO_3(s) + H_2O(l) + CO_2(g)$$

(v) in the final stage the ammonia is recovered from the ammonium chloride:

$$2NH_4Cl(aq) + CaO(s) \rightarrow$$

$$CaCl_2(aq) + 2NH_3(g) + H_2O(l)$$

The ammonia and carbon dioxide produced in these last two stages are recycled and used again. The only **by-product** of the reaction is calcium chloride ($CaCl_2$).

solvent A liquid in which a **solute** dissolves to form a **solution**. Water is the most common example. Solvents may be **polar**, such as water, or non-polar, for example **ether**. Polar solvents dissolve ionic or polar solutes, such as **salts**. Non-polar solvents dissolve **covalent compounds**, such as **hydrocarbons**.

specific heat capacity The amount of **heat energy** needed to increase the temperature of 1 g of a substance or material by 1 degree of temperature. The value for water is 4.2 joules per gram, per kelvin.

spectator ions **Ions** that play no part in a reaction. They are only 'spectators'. In the example shown below the spectator ions are underlined.

$$\underline{Na^+}(aq) + \underline{Cl^-}(aq) + Ag^+(aq) + \underline{NO_3^-}(aq) \rightarrow \underline{Na^+}(aq) + \underline{NO_3^-}(aq) + AgCl(s)$$

The **ionic equation** for this reaction is:

$$Ag^+(aq) + Cl^-(aq) \rightarrow AgCl(s)$$

stainless steel *See* **steel**.

stalactites and stalagmites Mineral growths found in **limestone** caves. When water containing calcium salts drips into the cave there is a gradual build-up of calcium carbonate from the roof (forming stalactites) and onto the floor (forming stalagmites) as the solution containing the calcium salts decomposes:

$$Ca(HCO_3)_2(aq) \rightarrow CaCO_3(s) + H_2O(l) + CO_2(g)$$

starch (($C_6H_{10}O_5)_n$) A **carbohydrate** polymer. It is made up of **monomers** of **glucose** and can be broken down by the action of **enzymes** or dilute **acid**. Glucose or **maltose** is formed.

Starch is a white tasteless powder that is insoluble in water. It is produced in most plants as a food store; plants that are especially rich in starch include cereals, legumes and tubers, such as potatoes. It is a major energy source for animals who eat the plants.

Iodine solution turns blue in the presence of starch; this can be used as a test for either iodine or starch. *See also* **polysaccharide**.

states of matter   The physical form in which substances exist. There are three states of matter: **solid**, **liquid** and **gas**. Substances are changed from one state into another by altering the temperature or changing the pressure. *See also* **change of state**.

state symbols   The letters placed next to the **formula** of a substance, usually in brackets, in a chemical **equation** to denote the **state of matter** of that substance in the reaction. They are: (s) = **solid**, (l) = **liquid**, (g) = **gas**, (aq) = **aqueous solution**.

steam   **Water** vapour at a temperature above its boiling point. Steam is a colourless gas. The 'steam' that is seen coming from a kettle is normally below the boiling point and is made up of droplets of water that have cooled and condensed. *See also* **condensation**.

steam reformation   The reaction of **methane** with steam (700°C, 30 atmospheres pressure, nickel **catalyst**) to form carbon monoxide and hydrogen:

$$CH_4(g) + H_2O(g) \rightarrow CO(g) + 3H_2(g)$$

See also **shift reaction, synthesis gas**.

steel   An **alloy** that contains iron as the main constituent. By altering the type and proportions of other elements in the alloy, substances of varying properties can be formed. The two best known steels are *mild steel*, which is used for car bodies and household goods such as cookers, freezers, etc., and *stainless steel*, which is used in industry and in cooking utensils. Mild steel **rusts** easily and so has to be protected by **galvanizing**, enamelling or painting. Stainless steel is not corroded by oxygen and so is very useful. The alloy contains the following proportions:

iron	74%
chromium	18%
nickel	8%

Steel-cutting tools contain tungsten, and high-temperature steels contain molybdenum. *See also* **steel manufacture**.

steel manufacture The industrial processes by which **steel** is made. There are two methods, the *basic oxygen process* and the *electric arc process.*

In the basic oxygen process, scrap steel (25% of the total mass) and a little **limestone** are dissolved in molten iron (which comes straight from the **blast furnace**). Pure oxygen is then blown onto the surface of the molten mixture and the impurities, such as carbon, are burnt off. Other metals are added to produce steel with the correct composition and, when the process is complete, the steel is cast into long continuous strips. The process is used to produce everyday steels used in heavy engineering, canning, car bodies and household goods.

In the electric arc process, the raw material is solid scrap steel. A large electric current is passed through the steel until it melts. Oxygen is used to burn off impurities, and additional materials are added to produce the required composition. The process is used to produce more specialist steels such as stainless and surgical steels.

sterilization A heat-treatment process to reduce the number of microorganisms present in a substance. This extends the shelf-life of food such as milk, which is sterilized in the bottle by being heated to between 115 °C and 130 °C for between 10 and 20 minutes. Sterilized milk can be kept without refrigeration for up to 5 months. Surgical and dental instruments are examples of tools that are kept free from microorganisms by sterilization. *See also* **pasteurization**, **UHT**.

STP (standard temperature and pressure) The standard conditions of one **atmosphere** pressure and a temperature of 0 °C (273 K). When comparing gas volumes it is useful to have standard conditions to refer to.

strength of acids and bases The degree of **dissociation** of an **acid** or a **base**. *Strong acids* and *strong bases* are fully dissociated in solution. Good examples are acids such as hydrochloric, nitric and sulphuric acid, and bases such as sodium or potassium hydroxide.

$$HCl(aq) \rightleftharpoons H^+(aq) + Cl^-(aq)$$
$$\text{pH range } 1\text{--}2$$
$$NaOH(aq) \rightleftharpoons Na^+(aq) + OH^-(aq)$$
$$\text{pH range } 13\text{--}14$$

Weak acids and *weak bases* do not fully dissociate in solution. Good examples are found in the organic acids, such as citric, tartaric, ethanoic and malic acids; and in bases such as ammonia.

$$CH_3COOH(aq) \rightleftharpoons CH_3COO^-(aq) + H^+(aq)$$
pH range 3–6

$$NH_3(aq) + H_2O(l) \rightleftharpoons NH_4^+(aq) + OH^-(aq)$$
pH range 8–11

strength of acids and bases

strong acid/base *See* **strength of acids and bases**.

structural formula A chemical **formula** that shows the chemical bonds between atoms and the position of the atoms with respect to each other. By contrast, the **molecular formula** only shows the number of atoms in the molecule. Structural formulae are usually shown as two-dimensional diagrams, even though the molecule may not be **planar**. Examples are:

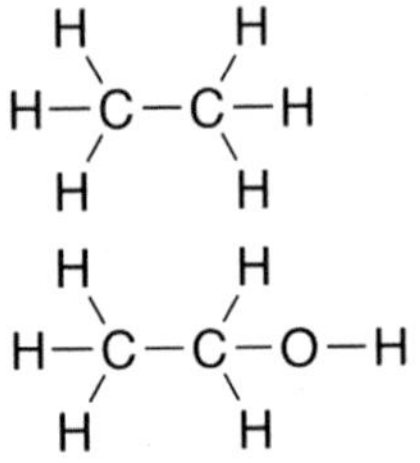

structural formula

subatomic particles Particles found in the atom. The important ones are the **proton**, **neutron** and **electron**.

sublimation The changing of a substance direct from the **solid** state to a **gas** without first melting. Common substances that *sublime* are carbon dioxide (CO_2) and iron(II) chloride ($FeCl_2$).

substitution reaction A reaction in which some atoms or groups in a molecule are replaced by others. It is a common reaction of **organic** compounds. For example:

$$CH_4(g) + Cl_2(g) \rightarrow CH_3Cl(l) + HCl(g)$$

A C–H bond is replaced by a C–Cl bond

sucrose ($C_{12}H_{22}O_{11}$) A **disaccharide** that is made up of a **fructose** unit and a **glucose** unit. It is the white crystalline **sugar** that is used in the home. Sucrose is obtained from sugar beet and cane sugar.

sugar The common general term for those sweet compounds which chemically are **monosaccharides** and **disaccharides**. Examples include

glucose, **sucrose**, **fructose** and **maltose**. The term also commonly refers to sucrose. *See also* **carbohydrate**.

sulphates   Compounds containing the SO_4^{2-} ion (**valency** = 2). They are widespread in nature, for example as **gypsum** ($CaSO_4$). They can be produced in the laboratory by the action of sulphuric acid on metals or **oxides** and **hydroxides**.

   The test for a sulphate is to add a solution to an acidified (HCl) solution of barium chloride. If a sulphate is present a white **precipitate** is formed that is insoluble in dilute acid.

sulphides   Compounds of sulphur with another element. They are produced by reaction with sulphur or hydrogen sulphide:

$$Fe(s) + S(s) \rightarrow FeS(s)$$

$$CuSO_4(aq) + H_2S(g) \rightarrow CuS(s) + H_2SO_4(aq)$$

sulphites   Compounds containing the SO_3^{2-} ion. They can be thought of as **salts** of **sulphurous acid** H_2SO_3. Sulphur(IV) oxide is prepared in the laboratory by the action of acid on a sulphite.

sulphur (S)   A non-metallic element that exists as two allotropic forms (*see* **allotrope**). *Rhombic sulphur* is a yellow solid that is stable below 96 °C, while the red solid *monoclinic sulphur* is stable at higher temperatures. The figures in the chart below refer to the monoclinic allotrope.

   Sulphur is in group VI of the **periodic table** and is reactive towards metals and oxygen. It is found uncombined in nature as well as occurring as metal sulphides, for example as galena (PbS) and pyrite (FeS).

   Sulphur-containing compounds are also found in **petroleum** and **natural gas**, which are now the major sources of the element. Sulphur is an important constituent of some drugs, such as the sulphonamides, and is used in the manufacture of **agrochemicals** and **dyes**, but its major use is in the manufacture of sulphuric acid in the **contact process**. *See also* **desulphurization**, **Frasch process**, **vulcanizing**.

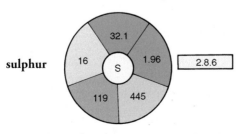

sulphur

sulphur compounds

| Sulphur(IV) oxide SO_2 | Commonly called sulphur dioxide, it is a colourless gas with a sharp odour. It is produced by burning sulphur or sulphur compounds in air or oxygen. It dissolves in water to form sulphurous acid – a weak acid. It is useful as a sterilizing agent and is often added to soft drinks. It is a reducing agent. This property is used in the chemical test for the gas. Sulphur dioxide is a major cause of air pollution and acid rain, being produced by the combustion of fossil fuels. |
| Sulphur(VI) oxide SO_3 | Sulphur trioxide: a volatile solid with a sharp odour. It is produced in the contact process by the combination of sulphur dioxide and oxygen over a catalyst (vanadium peroxide). The oxide is acidic and when it is added to water, sulphuric acid is formed. |

sulphuric acid (H_2SO_4)   A colourless, oily liquid that is a strong acid (*see* **strength of acids and bases**) and a vigorous **oxidizing agent**. It is one of the most important chemicals produced. It is made in the **contact process**. The chart opposite shows the main uses of the acid and the approximate amounts of the acid that are used.

Sulphuric acid reacts chemically in several ways. As an acid, dilute sulphuric acid reacts with metals, **bases** and **carbonates** to form sulphates:

$$Mg(s) + H_2SO_4(aq) \rightarrow MgSO_4(aq) + H_2(g)$$

$$CuO(s) + H_2SO_4(aq) \rightarrow CuSO_4(aq) + H_2O(l)$$

Concentrated sulphuric acid reacts with chlorides and nitrates to form hydrogen chloride and nitric acid respectively:

$$H_2SO_4(l) + NaCl(s) \rightarrow NaHSO_4(s) + HCl(g)$$

$$H_2SO_4(l) + NaNO_3(s) \rightarrow NaHSO_4(s) + HNO_3(g)$$

The concentrated acid is **hygroscopic** and acts as a dehydrating agent. Concentrated sulphuric acid is sometimes used to dry gases.

If the concentrated acid is added to **sucrose**, the sugar turns black. This is because water is removed from the **carbohydrate** and carbon is left behind:

$$C_6H_{12}O_6(s) \rightarrow 6C(s) + 6H_2O(g)$$

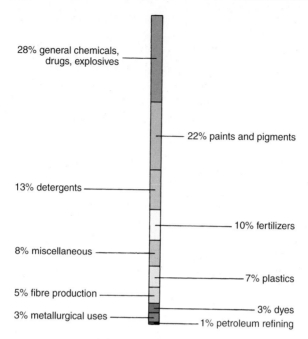

28% general chemicals, drugs, explosives

22% paints and pigments

13% detergents

10% fertilizers

8% miscellaneous

7% plastics

5% fibre production

3% dyes

3% metallurgical uses

1% petroleum refining

sulphuric acid   *Main uses.*

The reaction between concentrated suphuric acid and water is a very vigorous **exothermic reaction**. It is important to always add the acid to water and not the other way round.

Sulphuric acid also acts as an oxidizing agent. For example, although copper cannot displace hydrogen from acids, the metal can be oxidized (*see* **oxidation**) by concentrated sulphuric acid:

$$Cu(s) + 2H_2SO_4(l) \rightarrow CuSO_4(s) + SO_2(g) + 2H_2O(l)$$

sulphurous acid (H_2SO_3)   An acid that only exists in **aqueous solution**, formed by passing sulphur dioxide into water.

It is a weak acid (*see* **strength of acids and bases**) and a **reducing agent**. When heated, sulphur dioxide and steam are produced, leaving no residue. **Salts** of the acid are called **sulphites**.

supercooled liquid   *See* **glass**.

superphosphate   A **fertilizer** that is a valuable source of **phosphorus**. Superphosphate is formed by heating (insoluble) calcium phosphate and sulphuric acid under pressure. It is a mixture of calcium dihydrogen-phosphate, (which is **soluble**), calcium sulphate and other compounds.

supersaturated solution A **solution** that contains a higher **concentration** of **solute** than a **saturated solution**. It is usually produced by slowly cooling the saturated solution. If the solution is disturbed, for example by a mechanical shock, by dust falling into it or by a crystal of the solute being added to it, the excess solute will usually crystallize out. *See also* **crystallization**.

surfactant *See* **detergent**.

suspension Large particles (greater than 1000 nm) of insoluble solids or liquids suspended within a liquid. In time, the particles sink to the bottom of the vessel. Suspensions can be separated by use of a **filter** or a centrifuge. Blood is an example of a suspension, with the red and white blood cells suspended in the plasma. *See also* **colloidal dispersion, solution**.

symbol (in chemistry) One or more letters that are used to represent the names of substances – for example, Cl for chlorine. Symbols are also used in representing **units**.

synthesis The formation of a compound from smaller or simpler compounds or elements. For example the synthesis of ammonia from hydrogen and nitrogen. *See also* **synthesis gas**.

synthesis gas A mixture of carbon monoxide and hydrogen gas produced by the steam reformation of **natural gas** or **naphtha** using a nickel oxide catalyst. There are two steps in the process:

$$CH_4(g) + H_2O(g) \rightarrow 3H_2(g) + CO(g)$$

$$CO(g) + H_2O(g) \rightarrow H_2(g) + CO_2(g)$$

The second step is known as the **shift reaction**.
   As its name implies, synthesis gas is used to produce other chemicals. It is an important feedstock for the chemical industry. *See also* **methanol**.

synthetic Made from artificial rather than natural substances. For example, **nylon** is a synthetic **fibre**, whereas **silk** is a natural fibre.

T

tar Any of various heavy, thick liquids produced by the **distillation** (in the absence of air) of carbon-containing materials such as wood or coal. Tars are rich sources of **organic** compounds.

temperature A measure of the **kinetic energy** of a substance. The temperature scale used in everyday life is the Celsius (or centigrade) scale. In scientific work, the **Kelvin temperature scale** is usually used. *See also* **Celsius scale of temperature, room temperature**.

tempering The strengthening or toughening of a metal or metal object by heat treatment, usually by first heating it and then quenching it in water or oil. *See also* **annealing**.

temporarily hard water *See* **hard water**.

terylene *See* **polyester**.

tetrachloromethane A dense, colourless liquid that is **immiscible** with water. It is a good **solvent** for non-**polar** compounds. It can be made from **methane** by a **substitution reaction** involving chlorine:

$$4Cl_2(g) + CH_4(g) \rightarrow CCl_4(l) + HCl(g)$$

Other substituted methane products are also formed (for example CH_3Cl, CH_2Cl_2, $CHCl_3$) and the mixture must be separated to obtain pure products.

tetrahedral molecule A molecule in which four atoms are arranged as if at the corners of a tetrahedron and linked to an atom at the centre of it by **covalent bonds** or **coordinate bonds**. In a carbon compound such as **methane**, where there are four **single bonds** leading to four atoms, the atoms are arranged tetrahedrally around the carbon atom. The atoms are as far away from each other as possible. The angle between the bonds is 109° 28′, the *tetrahedral angle*.

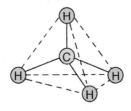

tetrahedral compound

thermal Relating to heat. For example thermal **decomposition** is decomposition by heat. Thermal energy is **heat energy**.

thermit reaction A reaction whereby aluminium removes the oxygen from the **oxide** of a metal that is below it in the **reactivity series**. Iron(II) oxide was used in industry to provide a small source of molten metal. Great heat is liberated in the reaction and the products are iron and aluminium oxide:

$$Fe_2O_3(s) + 2Al(s) \rightarrow Al_2O_3(s) + 2Fe(l)$$

thermometer An instrument for measuring temperature in order to show how hot something is. Thermometers come in different forms; the 'liquid in glass' thermometer is a common type. The liquid – most commonly mercury – is held in a bulb. As the temperature rises, the liquid expands in the bulb and rises up the tube. The tube is marked at intervals with the temperature and this can be read off.

thermoplastic polymer or thermosoftening polymer A **polymer** that softens when it is heated and which can therefore be moulded and remoulded into new shapes. Examples are **nylon** and **poly(chloroethene)**.

thermosetting polymer A **polymer** that cannot be softened again once it has been formed. Decomposition occurs if it is heated again. Examples are bakelite and formica.

tin (Sn) A metal that is in group IV of the **periodic table**. It occurs as **allotropes**, the most common being grey tin and white tin. The data given in the diagram refers to the white form. White tin is a shiny metal that shows normal metallic reactions. It is found naturally as the **oxide**, SnO_2. The grey allotrope is stable at low temperatures.

Tin is used to make **tinplate**, **bronze**, **solder** and **alloys** for bearings. It is also used in the production of **glass**: glass is floated onto the surface of molten tin, allowing large areas of glass without defects to be formed.

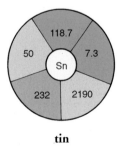

tin

tinplate Mild **steel** that, by **electroplating**, has been given a thin layer of **tin**. Tinplate is used in the canning industry to provide an **inert** layer between the steel of the can and the food.

titanium (Ti) A **transition metal** that is common in the Earth's crust (0.5%). It forms **alloys** that are very strong while having a low density. It is resistant to **corrosion** up to a high temperature. Titanium alloys are used in the aircraft industry. Titanium(IV) oxide TiO_2 is used as a **pigment**.

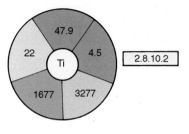

titanium

titration The reaction of two solutions used in **analysis** to discover the **concentration** of one of the solutions (the other concentration must be known). An accurately measured volume of a solution is normally placed in a conical **flask** and the other in a **burette**. The solution from the burette is added slowly until the solution in the flask has been completely used up by the reaction. This is shown by the use of an **indicator** or by means of a **pH** meter or a conductivity meter. The volume of solution used can then be read off the burette scale. The technique is most commonly used for **acid-base reactions**. *See also* **volumetric analysis**.

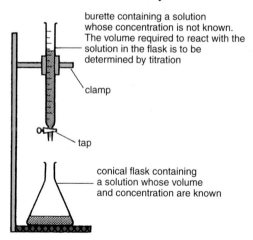

burette containing a solution whose concentration is not known. The volume required to react with the solution in the flask is to be determined by titration

clamp

tap

conical flask containing a solution whose volume and concentration are known

titration

toluene *See* **methylbenzene**.

tonne **or** metric ton A unit of **mass**. One tonne equals 1000 kilograms. It is equal to 2205 lb and so is 35 lb lighter than the imperial ton (2240 lb).

trace element An **element** that is needed in very small amounts in the body or by plants.

transition metals The elements found in the central section of the **periodic table**. There are three series of them but the most important are those elements from scandium to zinc. Transition metals all have similar characteristic properties:
(a) they produce coloured compounds;
(b) they have variable **valency**;
(c) they are solids with high melting and boiling points;
(d) they are useful as **catalysts**;
(e) they have **cations** that are useful as catalysts.
Transition elements tend to be important elements for use as pure metals, e.g. iron, copper, silver, gold, mercury, zinc, platinum, or for use in **alloys**, e.g. titanium, vanadium, chromium, manganese, cobalt, nickel, tungsten.

transition temperature The temperature above which one **allotrope** is stable and below which another is stable. Some examples are shown here:
(a) rhombic sulphur is stable below 96 °C, monoclinic sulphur is stable above 96 °C;
(b) grey tin is stable below 13 °C, white tin is stable above 13 °C.

trichloromethane **or** chloroform (CHCl$_3$) A substituted **alkane** that has been used as an anaesthetic. It is a colourless, volatile liquid that boils at 62 °C. *See also* **tetrachloromethane**.

triple bond A chemical **bond** that contains three shared pairs of electrons and is found in **alkynes** and nitrogen-containing compounds. Examples include:

$$\text{ethyne } H-C\equiv C-H \qquad \text{nitrogen } N\equiv N.$$

tritium The **isotope** of **hydrogen** that contains two **neutrons** in the atom. Tritium is a radioactive gas (*see* **radioactivity**).

U

UHT (ultra-heat treatment) A heat-treatment process to reduce the number of microorganisms present in food in order to extend its shelf life. UHT milk is heated to 140 °C for between 2 and 4 seconds and then cooled rapidly to 20 °C. If it is packed under sterile conditions and not opened, UHT milk can be kept at room temperature for up to 5 months. *See also* **pasteurization**, **sterilization**.

ultraviolet radiation (UV) Invisible radiation that has a slightly higher frequency (energy) than violet light. It is produced in large amounts by stars, e.g. the Sun. Long exposure to intense ultraviolet radiation is harmful to humans – skin cancers can result. Not much of the UV radiation from the Sun reaches the Earth, but is instead filtered out by the **ozone layer** in the stratosphere. *See also* **Earth's atmosphere**.

units Fixed quantities that are used as standards to measure other things. *See also* **SI units**.

Measurement	Unit
Distance	Metre (m)
Time	Second (s)
Mass	Kilogram (kg)

units

universal indicator A mixture of indicators. Because it is a mixture it changes colour several times as the **pH** of the solution changes. It is possible to tell the approximate pH of a solution by adding a few drops of universal indicator to it and reading the pH off a coloured chart. A summary of the colours is shown below. Universal indicator paper strips are also available.

Colour ←red orange yellow green				blue purple→			
pH	0 1 2 3 4 5 6 7 8 9	10	11	12	13	14	

universal indicator

unleaded petrol **Petrol** that does not contain the fuel additive tetraethyl lead. Instead, **methanol** (5%) and its derivative, *methyl tertiary butyl ether* (MTBE) (15%), are used. Unleaded fuel does not give rise to lead **pollution**, but other pollutants are still produced. *See also* **octane rating**.

unsaturated compounds Carbon compounds that possess **double bonds** or **triple bonds** between two carbon atoms. They contain more electrons in their bond than a normal single bond. Because of this they are reactive. Unsaturated compounds tend to react through **addition reactions,** forming new bonds with the electrons they possess; for example, **alkenes** become **alkanes.** *See also* **hydrocarbon, saturated compounds.**

(a)
$$\ce{C=C}$$
alkenes

$$-C\equiv C-$$
alkynes

(b)
ethene (unsaturated) $\xrightarrow{H_2}$ ethane (saturated)

unsaturated compounds (a) *Common examples of unsaturated compounds,* (b) *addition reactions form new bonds.*

uranium (U) A metallic element that has three naturally occurring **isotopes** (masses 234 (trace), 235 (0.7%) and 238 (99.3%)). They are all radioactive and eventually decay to give stable isotopes of lead.

238.1

92   U   19.0

1133   3930

uranium

Uranium is used as a **fuel** in nuclear power stations.
In some types of reactor, uranium oxide U_3O_8 is used. In others, uranium containing a higher than normal proportion of uranium-235 is used (enriched uranium). In the fast-breeder reactor, uranium-238 is turned into plutonium-239, which can then be used as a fuel. *See also* **nuclear reaction.**

urea A colourless substance found in the urine of all mammals, but also produced commercially by the reaction of carbon dioxide and ammonia at 200°C and 400 atmospheres pressure.

urea

$$CO_2(g) + 2NH_3(g) \rightarrow NH_2CONH_2(g) + H_2O(g)$$

Urea is used as a **fertilizer,** in the production of **adhesives** and pharmaceuticals and in the production of **urea/methanal resins.**

urea/methanal resin A **condensation polymer** formed by the reaction of **urea**, NH_2CONH_2, and **methanal**, HCHO. It is used as an **adhesive** and in the production of electrical plugs and sockets. *See also* **melamine**.

urethanes **Monomers** for the production of polyurethanes, which are important in the making of synthetic **foam**.

V

vacuum A space where there are no atoms or molecules present. It is impossible to obtain a perfect vacuum but we talk of a *partial vacuum* when the pressure is extremely low. The pressure is low outside the Earth's atmosphere, in space, and this is usually referred to as a vacuum.

valency A *combining power*, i.e. the usual number of chemical **bonds** that an atom forms when making compounds. More precisely, the valency of an element is the number of electrons that it needs to form a compound or **radical**; this usually leaves the atom with a stable (noble gas) configuration of electrons.

The electrons can be given to another element; they can be taken from an element; or they can be shared.

Some elements always have the same valency, for example hydrogen = 1, oxygen = 2 (except in peroxides), sodium = 1, magnesium = 2. Some examples of how valency determines the types of molecules formed by an atom:

(a) Sodium has a valency of 1. Sodium gives one electron away when it forms the sodium ion Na^+, for example in NaCl.

(b) Oxygen has a valency of 2. Oxygen accepts two electrons when it forms the oxide ion O^{2-} or it forms **covalent compounds**, such as SO_2, CO_2, N_2O.

(c) Hydrogen and bromine have valency of 1. They both donate one electron to make the H–Br **covalent bond** when they form hydrogen bromide.

Transition elements, however, have more than one valency: iron = 2, 3; cobalt = 2, 3; copper = 1, 2.

The valencies of some common elements and ions are shown in the following tables:

+1		+2		+3	
Lithium	Li^+	Calcium	Ca^{2+}	Aluminium	Al^{3+}
Sodium	Na^+	Magnesium	Mg^{2+}	Iron(III)	Fe^{3+}
Potassium	K^+	Zinc	Zn^{2+}		
Silver	Ag^+	Iron(II)	Fe^{2+}		
Ammonium	NH_4^+	Lead	Pb^{2+}		

valency

-1		-2		-3	
Fluoride	F^-	Sulphide	S^{2-}	Phosphate	PO_4^{3-}
Chloride	Cl^-	Oxide	O^{2-}		
Bromide	Br^-	Carbonate	CO_3^{2-}		
Iodide	I^-	Sulphate	SO_4^{2-}		

valency

vanadium (V) A **transition metal**. Its chief use is in the production of vanadium-steel **alloys**, which are valuable because of their high tensile strength and hardness; that is, their ability to withstand stress. The alloys often include chromium or manganese in addition to vanadium.

Vanadium(V) oxide V_2O_5 is the **catalyst** used in the **contact process**.

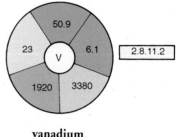

vanadium

van der Waals' forces or Loudon forces Weak forces between atoms. They are caused by the movement of electrons within the atoms. Elements that have only these forces to hold them together, such as the **noble gases**, have very low melting and boiling points.

vaporization The process by which a liquid becomes a **vapour**.

vapour Atoms or molecules that are in the gaseous state but below their *critical temperature*, i.e. the temperature above which liquid cannot exist.

For example, when water evaporates from a saucer left in the house, water vapour is formed and not **steam**.

vinegar A solution that is made by the action of bacteria on wine or cider. It contains about 4% **ethanoic acid**. It is used widely in the food industry for preserving foods.

vitamins Chemicals that are important to the proper working of the body. They tend to be complex **organic** molecules that cannot be made in the body but which must be eaten. Examples are vitamin A (found in e.g. dairy products) and vitamin C (found especially in citrus fruit).

volatile Easily turned into a **vapour**. Volatile substances either have boiling points that are near **room temperature** (examples are **ether** and propanone); or they are solids, such as carbon dioxide, which sublime (*see* **sublimation**). Liquids that are both volatile and flammable are very dangerous because of the risk of explosion. Care has to be taken with their storage and handling. Petrol is such a liquid.

volume The amount of space occupied by a substance. Solids and liquids have fixed volumes but a gas will have the same volume as the container it occupies. The larger the container is, the lower is the pressure for a fixed mass of gas. The volume of a gas can easily be changed by compressing it but it is much more difficult to change the volumes of solids and liquids. Hydraulic brakes in a car depend on this property of a liquid. Volumes are measured in cubic centimetres (cm^3).

volumetric analysis A method of **quantitative analysis** that uses accurately measured volumes of **solutions**. *See also* **burette**, **pipette**, **titration**.

vulcanization The process of adding **sulphur** to **rubber** to make it harder. *See also* **cross-linking**.

W

washing soda Hydrated sodium carbonate ($Na_2CO_3.10H_2O$). The name comes from the use of the salt to soften water that was to be used for washing. *See also* **hardness of water**.

water (H_2O) An **oxide** of **hydrogen**:

$$2H_2(g) + O_2(g) \rightarrow 2H_2O(l)$$

It is one of the most common compounds on Earth. It is the best known **solvent** and is needed by all living things.

Water is a colourless liquid and some of its more important properties are a follows:
(a) It has a freezing point of 0 °C.
(b) It has a boiling point of 100 °C.
(c) It has a density of 1.0 g cm^{-3} (water has a maximum density at 4 °C). Unlike most other substances, water expands on solidification. This accounts for ice floating and water pipes bursting. It does not conduct electricity although it can be electrolysed (*see* **electrolysis**) if small amounts of acid or alkali are added. The products are hydrogen and oxygen:

$$2H_2O(l) \rightarrow 2H_2(g) + O_2(g)$$

A test for the presence of water is the change in colour of **anhydrous salts**:

$$CuSO_4(s) + 5H_2O \rightarrow CuSO_4.5H_2O(s)$$
$$\text{white} \qquad\qquad\quad \text{blue}$$

These tests show that water is present but not that the water is **pure**. To show whether or not it is pure water, the boiling point of the liquid could be taken.

Water is abundant in the Earth's atmosphere, in lakes, rivers, glaciers and the oceans. It is found in rocks and in living creatures. Water is continually moving from place to place on the Earth (*see* **water cycle**).

The most important chemical property of water is its use as a **solvent**. Water has **polar bonds** and so can dissolve ionic solids such as sodium chloride (NaCl) as well as polar solids such as **glucose** ($C_6H_{12}O_6$).

The water which we drink is never pure. It always contains small amounts of dissolved gas (such as oxygen and carbon dioxide) and, depending on the source of the water, solids are dissolved in it too, some of which make the water hard (*see* **hard water**).

water cycle The continual movement of water around the Earth, in the oceans, on land and in the atmosphere. Water falls to Earth as rain, snow, hail or sleet, and freezes out of the air as frost and ice. It falls onto the oceans and these act as a vast source of water. It falls on land and here it enters the Earth and eventually flows into lakes and rivers and thus into the oceans. From these large areas of water, evaporation occurs and water re-enters the atmosphere. In this way the cycle continues. Plants and animals take water out of the ground for their own use and it can re-enter the atmosphere (by transpiration, excretion and **respiration**) or the Earth. Large quantities of water are stored as glaciers and the polar ice caps.

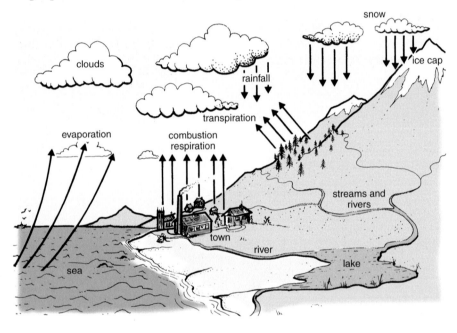

water cycle

water of crystallization **Water** that is chemically bonded within **crystals**. Examples are:

$$CuSO_4.5H_2O \quad Na_2CO_3.10H_2O \quad CaSO_4.2H_2O$$

Water can be removed by heating, leaving the **anhydrous** salt. This can happen because the chemical bonds holding the water molecules to those of the **salt** are weak. Some salts that include water of crystallization lose some of the water simply on exposure to the air, resulting in **efflorescence**.

water treatment *See* **sewage treatment**.

weak acids and bases *See* **strength of acids and bases**.

whitewash A **suspension** of calcium hydroxide in water. This is used as a temporary '**paint**' for things such as the white lines on sports pitches. Exposure to carbon dioxide in the air turns the slightly soluble hydroxide into the insoluble calcium carbonate:

$$Ca(OH)_2(aq) + CO_2(g) \rightarrow CaCO_3(s) + H_2O(g)$$

wool A natural **fibre** that comes from sheep, goats and similar animals. Wool is a **protein** fibre that has a crinkled and scaly structure. Because of this, the fibres are elastic and can trap air. Clothes made from wool tend to be warm.

wrought iron Iron which contains less than 0.25% carbon. *See also* **cast iron**, **steel**, **steel manufacture**.

X – Y – Z

xenon (Xe) One of the few **noble gases** that have been found to form compounds. Even so, it only binds with the two most reactive elements, fluorine and oxygen, for example in the compounds XeF_2, XeF_4, XeF_6, XeO_3. The element is used to fill light bulbs and fluorescent tubes.

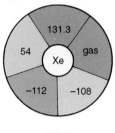

xenon

X-rays High-energy electromagnetic radiation. It is produced by firing electrons at metals. It is a very penetrating radiation and will easily pass through flesh but is stopped by bone and other dense substances such as metals.

yeasts Microscopic organisms that are very useful to humans. They are used in baking and brewing. In brewing, they are used to convert **sugars** into **ethanol**:

$$C_6H_{12}O_6(aq) \quad \xrightarrow{\text{yeast}} \quad 2C_2H_5OH(aq) \quad + \quad 2CO_2(g)$$

glucose   **ethanol**   **carbon dioxide**

The results are alcoholic drinks such as beer and wine. The carbon dioxide is either collected and sold or allowed to escape into the atmosphere.

In baking, it is the carbon dioxide that is useful. The dough is mixed and then the sugar and yeast react in the dough to produce the gas. The gas makes the dough expand (rise) and the mixture becomes much lighter. Yeasts contain **enzymes** and it is these substances that act on the sugars. *See also* **fermentation, zymase**.

yield (in chemistry) The amount of product that a chemical **reaction** produces. Many reactions do not produce as much product as would be expected from looking at the chemical equations. It is useful to express the amount of product as a percentage of what it is theoretically possible to produce:

$$\text{Yield of reaction} = \frac{\text{amount of product produced}}{\text{maximum amount of product that it is possible to produce}}$$

For example:

$$CuO(s) + H_2(g) \rightarrow Cu(s) + H_2O$$
copper(II)   hydrogen   copper   water
oxide
80 g $\rightarrow$ 64 g (maximum yeld)

The M_r of CuO is 80 and the A_r of Cu is 64. Therefore 80 g of CuO would give a maximum yield of 64 g of Cu. However, if 80 g of oxide actually produce 56 g of copper metal the percentage yield of the reaction is $(^{56}/_{64} \times 100)\% = 87.5\%$.

Z The symbol that is given to represent the **atomic number** of an element, i.e. the number of **protons** in the atom.

zinc (Zn) A **transition metal** that is found in the first transition series in the **periodic table**.

It is a dense, grey metal that is reactive towards acids but which does not react with cold water. The metal is used in the alloy **brass** and also in the protection of steel by **galvanizing**. Zinc is extracted from the sulphide ore (ZnS).

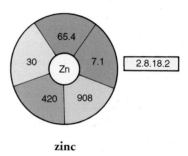

zinc

zinc compounds

Zinc oxide   This has a use in medicine – zinc oxide cream.
ZnO   It is used as a protection against skin irritations, such as nappy rash. It is also used in paints. Zinc oxide and hydroxide are amphoteric.

zymase The enzyme present in **yeasts** which is responsible for the formation of **ethanol** and carbon dioxide from **sugars**.

Appendix A

A list of some useful common abbreviations and symbols you may encounter in scientific literature.

A	(1) mass number
	(2) ampere – unit of electric current
aq	state symbol for aqueous solution, usually as (aq)
$\mathbf{A_r}$	relative atomic mass
atm	atmosphere – a unit of pressure
α	alpha (Greek letter)
ß	beta (Greek letter)
bp	boiling point
C	coulomb – unit of electric charge (quantity of electricity)
°C	degree Celsius
$\mathbf{cm^3}$	cubic centimetre, unit of volume
DC	direct current – the type of electricity produced from a battery
$\mathbf{dm^3}$	cubic decimetre $\equiv$ litre, unit of volume
E	symbol for emf of a cell
e or $\mathbf{e^-}$	electron
emf	electromotive force
g	(1) gram – unit of mass
	(2) state symbol for gas, usually as (g)
	(3) acceleration due to gravity
H	enthalpy (ΔH = enthalpy change)
I	electric current
IR	infrared radiation
J	joule – unit of energy
k	prefix meaning 'one thousand times' i.e. kg = 1000 g
K	kelvin – unit of temperature; the unit is the same as that of the Celsius scale i.e. a kelvin is equal to a degree Celsius
l	state symbol for liquid, usually as (l)

m	(1) mass
	(2) metre – unit of length
M	molar – unit of concentration (molarity) e.g. 2 M
m^3	cubic metre – unit of volume
mol	mole – unit of amount of substance
ml	millilitre, 1/1000 of 1 litre $\equiv$ 1 cm^3
mp	melting point
M_r	relative molecular mass
n	neutron
N	newton – unit of force
nm	nanometre, 10^{-9} m
NTP	normal temperature and pressure
p	(1) proton
	(2) pressure
Pa	pascal – unit of pressure
pd	potential difference
pH	relates to a scale of acidity, e.g. pH = 1: very strongly acidic
Q	electric charge (quantity of electricity)
s	(1) state symbol for solid, usually as (s)
	(2) second – unit of time
STP	standard temperature and pressure
t	time
T	temperature
$T^1/_2$ or $t^1/_2$	half-life (of radioactive isotope)
UV	ultraviolet
V	(1) volume
	(2) electrical potential difference (pd)
	(3) volt – unit of pd
Z	atomic number

Appendix B

Some common hazard signs and their meanings.

EXPLOSIVE
This substance may explode if ignited, heated, or exposed to friction or a sudden shock.

IRRITANT
This substance causes irritation to living tissue, e.g. skin may become red or blister after repeated contact.

OXIDIZING
This substance can cause fire when in contact with combustible material.

TOXIC
This substance is a serious health risk. Toxic effects may result from swallowing, inhalation or skin absorption.

HIGHLY FLAMMABLE
This substance may easily catch fire under normal laboratory conditions.

HARMFUL
This substance is less of a health risk than a TOXIC one but should still be handled with care. It may cause harm by swallowing, inhalation or skin absorption.

CORROSIVE
This substance can destroy living tissue.

RADIOACTIVE
radioactivity and should be treated with extreme care.